T0192795

SpringerBriefs in Statistics

JSS Research Series in Statistics

Editors-in-Chief

Naoto Kunitomo
Akimichi Takemura

Series editors

Genshiro Kitagawa
Tomoyuki Higuchi
Toshimitsu Hamasaki
Shigeyuki Matsui
Manabu Iwasaki
Yasuhiro Omori
Masafumi Akahira
Takahiro Hoshino
Masanobu Taniguchi

The current research of statistics in Japan has expanded in several directions in line with recent trends in academic activities in the area of statistics and statistical sciences over the globe. The core of these research activities in statistics in Japan has been the Japan Statistical Society (JSS). This society, the oldest and largest academic organization for statistics in Japan, was founded in 1931 by a handful of pioneer statisticians and economists and now has a history of about 80 years. Many distinguished scholars have been members, including the influential statistician Hirotugu Akaike, who was a past president of JSS, and the notable mathematician Kiyosi Itô, who was an earlier member of the Institute of Statistical Mathematics (ISM), which has been a closely related organization since the establishment of ISM. The society has two academic journals: the Journal of the Japan Statistical Society (English Series) and the Journal of the Japan Statistical Society (Japanese Series). The membership of JSS consists of researchers, teachers, and professional statisticians in many different fields including mathematics, statistics, engineering, medical sciences, government statistics, economics, business, psychology, education, and many other natural, biological, and social sciences.

The JSS Series of Statistics aims to publish recent results of current research activities in the areas of statistics and statistical sciences in Japan that otherwise would not be available in English; they are complementary to the two JSS academic journals, both English and Japanese. Because the scope of a research paper in academic journals inevitably has become narrowly focused and condensed in recent years, this series is intended to fill the gap between academic research activities and the form of a single academic paper.

The series will be of great interest to a wide audience of researchers, teachers, professional statisticians, and graduate students in many countries who are interested in statistics and statistical sciences, in statistical theory, and in various areas of statistical applications.

More information about this series at http://www.springer.com/series/13497

Ikuko Funatogawa · Takashi Funatogawa

Longitudinal Data Analysis

Autoregressive Linear Mixed Effects Models

 Springer

Ikuko Funatogawa
Department of Statistical Data Science
The Institute of Statistical Mathematics
Tachikawa, Tokyo, Japan

Takashi Funatogawa
Clinical Science and Strategy Department
Chugai Pharmaceutical Co. Ltd.
Chūō, Tokyo, Japan

ISSN 2191-544X   ISSN 2191-5458   (electronic)
SpringerBriefs in Statistics
ISSN 2364-0057   ISSN 2364-0065   (electronic)
JSS Research Series in Statistics
ISBN 978-981-10-0076-8   ISBN 978-981-10-0077-5   (eBook)
https://doi.org/10.1007/978-981-10-0077-5

Library of Congress Control Number: 2018960732

This Springer imprint is published by the registered company Springer Nature Singapore Pte Ltd.
The registered company address is: 152 Beach Road, #21-01/04 Gateway East, Singapore 189721, Singapore

Preface

There are already many books on longitudinal data analysis. This book is unique among them in that it specializes in autoregressive linear mixed effects models that we proposed. This is a new analytical approach for dynamic data repeatedly measured from multiple subjects over time. Random effects account for differences across subjects. Autoregression in the response itself is often used in time series analysis. In longitudinal data analysis, a static mixed effects model is changed into a dynamic one by the introduction of the autoregression term. It provides one of the simplest models that take into account the past covariate history without approximation and discrepancy between marginal and subject specific interpretation. Response levels in this model gradually move toward an asymptote or equilibrium which depends on fixed effects and random effects, and this is an intuitive summary measure. Linear mixed effects models have good properties but are not always satisfactory to express those nonlinear time trends. Little is known about what autoregressive linear mixed effects models represents when used in longitudinal data analysis.

Chapter 1 introduces longitudinal data, linear mixed effects models, and marginal models before the main theme. Prior knowledge of regression analysis and matrix calculation is desirable. Chapter 2 introduces autoregressive linear mixed effects models, the main theme of this book. Chapter 3 presents two case studies of actual data analysis about the topics of response-dependent dropouts and response-dependent dose modifications. Chapter 4 describes the bivariate extension, along with an example of actual data analysis. Chapter 5 explains the relationships with nonlinear mixed effects models, growth curves, and differential equations. Chapter 6 describes state space representation as an advanced topic for interested readers.

Our experiences with data analysis are mainly through experimental studies, such as randomized clinical trials and clinical studies with dose modifications. Here, we focus on the mechanistic aspects of autoregressive linear mixed effects models. Dynamic panel data analysis in economics is closely related to autoregressive linear mixed effects models; however, it is used in observational studies and is not covered in this book.

We would like to thank Professor Nan Laird, Professor Daniel Heitjan, Dr. Motoko Yoshihara, Dr. Keiichi Fukaya, and Dr Kazem Nasserinejad for reviewing the draft of this book. We believe that the book has been improved greatly. As the authors, we take full responsibility for the material in this book. This work is supported by JSPS KAKENHI Grant Number JP17K00066 and the ISM Cooperative Research Program (2018–ISMCRP–2044) to Ikuko Funatogawa. We would like to thank the professors in the institute of statistical mathematics who encouraged us for writing the book. We wrote the doctoral theses based on the autoregressive linear mixed effects models more than a decade ago. We are very grateful to Ikuko's supervisor, Professor Yasuo Ohashi, and Takashi's supervisor, Professor Masahiro Takeuchi. We would also like to thank Professor Manabu Iwasaki for giving us this valuable opportunity to write this book. Lastly, we extend our sincere thanks to our family.

Tokyo, Japan
December 2018

Ikuko Funatogawa
Takashi Funatogawa

Contents

Chapter 1
Longitudinal Data and Linear Mixed Effects Models

Abstract Longitudinal data are measurements or observations taken from multiple subjects repeatedly over time. The main theme of this book is to describe autoregressive linear mixed effects models for longitudinal data analysis. This model is an extension of linear mixed effects models and autoregressive models. This chapter introduces longitudinal data and linear mixed effects models before the main theme in the following chapters. Linear mixed effects models are popularly used for the analysis of longitudinal data of a continuous response variable. They are an extension of linear models by including random effects and variance covariance structures for random errors. Marginal models, which do not include random effects, are also introduced in the same framework. This chapter explains examples of popular linear mixed effects models and marginal models: means at each time point with a random intercept, means at each time point with an unstructured variance covariance, and linear time trend models with a random intercept and a random slope. The corresponding examples of group comparisons are also provided. This chapter also discusses the details of mean structures and variance covariance structures and provides estimation methods based on maximum likelihood and restricted maximum likelihood.

Keywords Linear mixed effects model · Longitudinal · Maximum likelihood Random effect · Unbalanced data

1.1 Longitudinal Data

Longitudinal data are measurements or observations taken from multiple subjects repeatedly over time. If multiple response variables are measured, the data are called multivariate longitudinal data. There are several interests to perform longitudinal data analysis, including changes in the response variable over time, differences in changes by factor or covariate, and relationships between changes in multiple response variables. Because measurements from the same subjects are not independent, analytical methods that take account of the correlation or variance covariance between repeated measures have been developed. For a continuous response variable, linear mixed effects models are often applied, and this model treats differences across subjects as random effects. In the 1980s, seminal papers about linear mixed effects models were published, such as that by Laird and Ware (1982). Since the 1990s, books about longitudinal data analysis have been published and statistical software has been developed. This book discusses the analysis of continuous response variables.

Analytical methods differ depending on the study designs. This can be illustrated by three study designs with 100 blood pressure measurements. If the data were measured once from 100 subjects, analytical methods such as t test or regression analysis are used. If the data were measured 100 times from one subject, analytical methods for time series data are used. If the data were repeatedly measured five times from 20 subjects, analytical methods for longitudinal data are used.

Figure 1.1a shows an example of longitudinal data which are hypothetical data from a randomized controlled trial (RCT). The solid and dotted lines indicate the changes in a response variable for each subject in two groups. In an RCT, groups such as a new treatment and placebo are randomly assigned to each subject. Then the distributions of the response variable at the randomization, baseline, are expected to be the same between the two groups. We compare the response variable after the randomization. The randomization guarantees comparability between groups, or internal validity. Especially, it is important that the distributions of unknown confounders will be the same between groups. This is a strong point of RCTs compared with observational studies. In observational studies, adjustment of confounding is important and we cannot adjust unknown confounders. This book covers RCTs and experimental studies rather than observational studies.

Figure 1.1b shows an example of time series data. Time series data usually have a large number of time points. Typical longitudinal data have a large or moderate

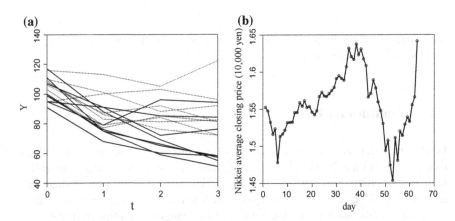

Fig. 1.1 **a** Hypothetical longitudinal data from a randomized controlled trial to compare two groups. Solid and dotted lines indicate changes in a response variable for each subject in two groups. **b** Time series data of Nikkei average closing price

number of subjects but a limited number of time points. There are also various cases of longitudinal data such as data with a large number of subjects and a large number of time points or data with a limited number of subjects and a large number of time points.

Since the 1990s, many books on longitudinal data analysis have been published. These include Diggle et al. (1994, 1st edn.), Diggle et al. (2002, 2nd edn.), Dwyer et al. (1992), Fitzmaurice et al. (2004, 1st edn.; 2011, 2nd edn.), Fitzmaurice et al. (2009), Gregoire et al. (1997), Hand and Crowder (1996), Jones (1993), Laird (2004), Littell et al. (1996, 1st edn.), Littell et al. (2006, 2nd edn.), Tango (2017), Verbeke and Molenberghs (1997, 2000), Vonesh (2012), Wu and Zhang (2006), and Zimmerman and Núñez-Antón (2010). The feature of this book is to discuss mainly autoregressive linear mixed effects models which are rarely introduced in other books.

Section 1.2 introduces linear mixed effects models and marginal models. Section 1.3 provides specific examples of these models. Section 1.4 shows mean structures and variance covariance structures generally used. Section 1.5 discusses the inference based on maximum likelihood.

1.2 Linear Mixed Effects Models

Linear mixed effects models are used for the analysis of longitudinal data of continuous response variables. Let $\mathbf{Y}_i = (Y_{i1}, Y_{i2}, \cdots, Y_{in_i})^T$ be the vector of the response corresponding to the ith ($i = 1, \cdots, N$) subject measured from 1 to n_i occasions. Y_{ij} is the jth measurement. $\mathbf{A}^T$ denotes the transpose of $\mathbf{A}$. Linear mixed effects models are expressed by

$$\mathbf{Y}_i = \mathbf{X}_i \boldsymbol{\beta} + \mathbf{Z}_i \mathbf{b}_i + \boldsymbol{\varepsilon}_i, \tag{1.2.1}$$

where $\boldsymbol{\beta}$ is a $p \times 1$ vector of unknown fixed effects parameters, $\mathbf{X}_i$ is a known $n_i \times p$ design matrix for fixed effects, $\mathbf{b}_i$ is a $q \times 1$ vector of unknown random effects parameters, $\mathbf{Z}_i$ is a known $n_i \times q$ design matrix for random effects, and $\boldsymbol{\varepsilon}_i$ is a $n_i \times 1$ vector of random errors, $\boldsymbol{\varepsilon}_i = (\varepsilon_{i1}, \varepsilon_{i2}, \cdots, \varepsilon_{in_i})^T$. It is assumed that $\mathbf{b}_i$ and $\boldsymbol{\varepsilon}_i$ are both independent across subjects and independently follow a multivariate normal distribution with the mean zero vector, $\mathbf{0}$, and variance covariance matrices $\mathbf{G}$ and $\mathbf{R}_i$, respectively. The distributions are expressed by

$$\mathbf{b}_i \sim \text{MVN}(\mathbf{0}, \mathbf{G}), \tag{1.2.2}$$

$$\boldsymbol{\varepsilon}_i \sim \text{MVN}(\mathbf{0}, \mathbf{R}_i), \tag{1.2.3}$$

where $\mathbf{G}$ is a $q \times q$ square matrix and $\mathbf{R}_i$ is an $n_i \times n_i$ square matrix. In these matrices, the diagonal elements are variance and the non-diagonal elements are covariance. These matrices include unknown parameters and are assumed to be some structures. Responses from different subjects are assumed to be independent.

Following the above assumptions, the marginal distribution of $\mathbf{Y}_i$ is a multivariate normal distribution. The mean vector is the marginal expectation $E(\mathbf{Y}_i) = \mathbf{X}_i\boldsymbol{\beta}$, and the variance covariance matrix $\mathbf{V}_i = \text{Var}(\mathbf{Y}_i)$ is

$$\mathbf{V}_i = \text{Var}(\mathbf{Z}_i\mathbf{b}_i + \boldsymbol{\varepsilon}_i) = \mathbf{Z}_i\mathbf{G}\mathbf{Z}_i^T + \mathbf{R}_i. \tag{1.2.4}$$

The distribution is expressed by

$$\mathbf{Y}_i \sim \text{MVN}(\mathbf{X}_i\boldsymbol{\beta}, \mathbf{V}_i), \tag{1.2.5}$$

where $\mathbf{V}_i$ is an $n_i \times n_i$ square matrix. The $E(\mathbf{Y}_i)$ and the expectation for a typical subject with $\mathbf{b}_i = \mathbf{0}$, $E(\mathbf{Y}_i|\mathbf{b}_i = \mathbf{0})$, are the same.

Correlations and unequal variances across responses are now allowed. Inclusion of random effects and structures of a variance covariance matrix are extensions of linear models. In linear models, the variance covariance matrix is usually assumed to be an independent structure with equal variances.

The term "mixed effects" means including both fixed effects and random effects. However, the models that assume some particular structures on $\mathbf{R}_i$ without random effects, $\mathbf{V}_i = \mathbf{R}_i$ instead of $\mathbf{V}_i = \mathbf{Z}_i\mathbf{G}\mathbf{Z}_i^T + \mathbf{R}_i$, are also discussed in the framework of linear mixed effects models. The models are

$$\mathbf{Y}_i = \mathbf{X}_i\boldsymbol{\beta} + \boldsymbol{\varepsilon}_i, \tag{1.2.6}$$

where $\boldsymbol{\varepsilon}_i \sim \text{MVN}(\mathbf{0}, \mathbf{R}_i)$, $E(\mathbf{Y}_i) = \mathbf{X}_i\boldsymbol{\beta}$, and $\text{Var}(\mathbf{Y}_i) = \mathbf{R}_i$. They are called marginal models. In marginal models, we model the marginal expectation of the response, $E(\mathbf{Y}_i)$, marginal variance, and correlation. We can define a multivariate normal distribution by the mean structure as the first moment and variance covariance structure as the second moment. Linear mixed effects models can be transformed to marginal models as (1.2.5). For discrete responses, such as binary and count data, however, higher moments are required to define likelihood, and the generalized estimating equation (GEE) is often used for marginal models. Furthermore, the interpretation of fixed effects parameters $\boldsymbol{\beta}$ differs between mixed effects models and marginal models.

1.3   Examples of Linear Mixed Effects Models

This section provides several specific linear mixed effects models and marginal models, which are simple and widely used. These include means at each time point with a random intercept, means at each time point with an unstructured (UN) variance covariance, and linear time trend models with a random intercept and a random slope. Examples of these models for group comparisons are also provided.

1.3.1 Means at Each Time Point with Random Intercept

In clinical trials and experimental studies, observation time points are often designed to be the same across subjects. Time intervals between observations are not necessary to be equal. If the observation time points are the same, we use the following model with means at each time point with a random intercept,

$$
\begin{cases}
Y_{ij} = \mu_j + b_i + \varepsilon_{ij} \\
b_i \sim N(0, \sigma_G^2) \\
\varepsilon_{ij} \sim N(0, \sigma_\varepsilon^2)
\end{cases}
\tag{1.3.1}
$$

Here, μ_j is the mean at the jth ($j = 1, \cdots, J$) time point. b_i is a random intercept for the ith subject. It is assumed that b_i and a random error, ε_{ij}, independently follow a normal distribution with the mean zero and the variances σ_G^2 and σ_ε^2, respectively.

In the case of four time points, the model for the response, $\mathbf{Y}_i$, and the variance covariance matrix of the response vector, $\mathbf{V}_i = \mathrm{Var}(\mathbf{Y}_i)$, are

$$
\mathbf{Y}_i = \begin{pmatrix} Y_{i1} \\ Y_{i2} \\ Y_{i3} \\ Y_{i4} \end{pmatrix} = \begin{pmatrix} 1 & 0 & 0 & 0 \\ 0 & 1 & 0 & 0 \\ 0 & 0 & 1 & 0 \\ 0 & 0 & 0 & 1 \end{pmatrix} \begin{pmatrix} \mu_1 \\ \mu_2 \\ \mu_3 \\ \mu_4 \end{pmatrix} + \begin{pmatrix} 1 \\ 1 \\ 1 \\ 1 \end{pmatrix} b_i + \begin{pmatrix} \varepsilon_{i1} \\ \varepsilon_{i2} \\ \varepsilon_{i3} \\ \varepsilon_{i4} \end{pmatrix},
$$

$$
\begin{aligned}
\mathbf{V}_i &= \mathbf{Z}_i \mathbf{G} \mathbf{Z}_i^T + \mathbf{R}_i \\
&= \mathbf{Z}_i \mathbf{G} \mathbf{Z}_i^T + \sigma_\varepsilon^2 \mathbf{I}_{n_i} \\
&= \begin{pmatrix} 1 \\ 1 \\ 1 \\ 1 \end{pmatrix} \sigma_G^2 (1\ 1\ 1\ 1) + \sigma_\varepsilon^2 \begin{pmatrix} 1 & 0 & 0 & 0 \\ 0 & 1 & 0 & 0 \\ 0 & 0 & 1 & 0 \\ 0 & 0 & 0 & 1 \end{pmatrix} \\
&= \begin{pmatrix} \sigma_G^2 & \sigma_G^2 & \sigma_G^2 & \sigma_G^2 \\ \sigma_G^2 & \sigma_G^2 & \sigma_G^2 & \sigma_G^2 \\ \sigma_G^2 & \sigma_G^2 & \sigma_G^2 & \sigma_G^2 \\ \sigma_G^2 & \sigma_G^2 & \sigma_G^2 & \sigma_G^2 \end{pmatrix} + \sigma_\varepsilon^2 \begin{pmatrix} 1 & 0 & 0 & 0 \\ 0 & 1 & 0 & 0 \\ 0 & 0 & 1 & 0 \\ 0 & 0 & 0 & 1 \end{pmatrix} \\
&= \begin{pmatrix} \sigma_G^2 + \sigma_\varepsilon^2 & \sigma_G^2 & \sigma_G^2 & \sigma_G^2 \\ \sigma_G^2 & \sigma_G^2 + \sigma_\varepsilon^2 & \sigma_G^2 & \sigma_G^2 \\ \sigma_G^2 & \sigma_G^2 & \sigma_G^2 + \sigma_\varepsilon^2 & \sigma_G^2 \\ \sigma_G^2 & \sigma_G^2 & \sigma_G^2 & \sigma_G^2 + \sigma_\varepsilon^2 \end{pmatrix}.
\end{aligned}
$$

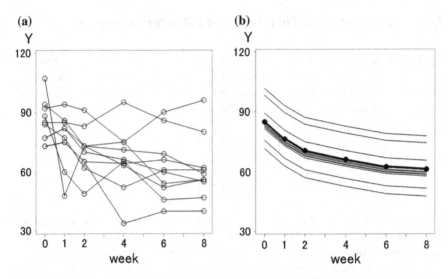

Fig. 1.2 **a** Data for selected subjects from schizophrenia trial data. **b** Means at each time point with a random intercept. Estimated means (thick line with closed circles) and predicted values for each subject (thin lines)

Here, $\mathbf{I}_a$ denotes an $a \times a$ identity matrix. The variance covariance structure, $\mathbf{Z}_i \mathbf{GZ}_i^T$, induced by a random intercept, is a square matrix with all the same elements, σ_G^2. The variance covariance matrix of the random error vector, $\sigma_\varepsilon^2 \mathbf{I}_{n_i}$, is added on to this matrix for $\mathbf{V}_i = \mathrm{Var}(\mathbf{Y}_i)$. The diagonal elements of $\mathbf{V}_i$ are $\sigma_G^2 + \sigma_\varepsilon^2$ and the non-diagonal elements are σ_G^2. σ_G^2 is called between-subject variance, inter-subject variance, or inter-individual variance. σ_ε^2 is called within-subject variance, intra-subject variance, or intra-individual variance.

Figure 1.2a shows longitudinal data for selected subjects from schizophrenia trial data in Diggle et al. (2002). We analyze the data in Chap. 3. Figure 1.2b shows an example using the model (1.3.1). The thick line with closed circles indicates the estimated means at each time point, $\hat{\mu}_j$. The thin lines show the predicted values, $\hat{\mu}_j + \hat{b}_i$, for each subject. The individual lines are parallel because a random intercept means the assumption of mutual parallelism across subjects over time.

The above model, $Y_{ij} = \mu_j + b_i + \varepsilon_{ij}$, is also expressed in another design matrix with $J - 1$ dummy variables, $x_{1j}, \cdots, x_{J-1j}$. Let $x_{kj} = 1$ if $k = j - 1$ and 0 otherwise. The model is

$$Y_{ij} = \beta_0 + \beta_1 x_{1j} + \cdots + \beta_{J-1} x_{J-1j} + b_i + \varepsilon_{ij}. \tag{1.3.2}$$

In the case of four time points, the model is

$$
\begin{pmatrix} Y_{i1} \\ Y_{i2} \\ Y_{i3} \\ Y_{i4} \end{pmatrix} = \begin{pmatrix} 1 & 0 & 0 & 0 \\ 1 & 1 & 0 & 0 \\ 1 & 0 & 1 & 0 \\ 1 & 0 & 0 & 1 \end{pmatrix} \begin{pmatrix} \beta_0 \\ \beta_1 \\ \beta_2 \\ \beta_3 \end{pmatrix} + \begin{pmatrix} 1 \\ 1 \\ 1 \\ 1 \end{pmatrix} b_i + \begin{pmatrix} \varepsilon_{i1} \\ \varepsilon_{i2} \\ \varepsilon_{i3} \\ \varepsilon_{i4} \end{pmatrix}.
$$

Data with the same observation time points without missing values are called balanced data. Data with different observation time points or data with missing values are called unbalanced data. Even if there are missing values, we can analyze all data without removing subjects with missing data. We discuss missing data in Sect. 3.2.

1.3.2 Group Comparison Based on Means at Each Time Point with Random Intercept

Here, we consider group comparisons based on the means at each time point with a random intercept. When there are two groups (A and B), let x_{gi} be an indicator variable for the group, with $x_{gi} = 0$ in group A and $x_{gi} = 1$ in group B. The linear mixed effects model with the main effects of time and the group with a random intercept is

$$
\begin{cases} Y_{ij} = \beta_0 + \beta_1 x_{1j} + \cdots + \beta_{J-1} x_{J-1j} + \beta_g x_{gi} + b_i + \varepsilon_{ij} \\ b_i \sim N(0, \sigma_G^2) \\ \varepsilon_{ij} \sim N(0, \sigma_\varepsilon^2) \end{cases} \tag{1.3.3}
$$

In the case of four time points, this model is

$$
\begin{pmatrix} Y_{i1} \\ Y_{i2} \\ Y_{i3} \\ Y_{i4} \end{pmatrix} = \begin{pmatrix} 1 & x_{11} & x_{21} & x_{31} & x_{gi} \\ 1 & x_{12} & x_{22} & x_{32} & x_{gi} \\ 1 & x_{13} & x_{23} & x_{33} & x_{gi} \\ 1 & x_{14} & x_{24} & x_{34} & x_{gi} \end{pmatrix} \begin{pmatrix} \beta_0 \\ \beta_1 \\ \beta_2 \\ \beta_3 \\ \beta_g \end{pmatrix} + \begin{pmatrix} 1 \\ 1 \\ 1 \\ 1 \end{pmatrix} b_i + \begin{pmatrix} \varepsilon_{i1} \\ \varepsilon_{i2} \\ \varepsilon_{i3} \\ \varepsilon_{i4} \end{pmatrix}.
$$

The design matrices of the fixed effects for groups A and B are

$$
\mathbf{X}_i = \begin{pmatrix} 1 & 0 & 0 & 0 & 0 \\ 1 & 1 & 0 & 0 & 0 \\ 1 & 0 & 1 & 0 & 0 \\ 1 & 0 & 0 & 1 & 0 \end{pmatrix} \text{ and } \mathbf{X}_i = \begin{pmatrix} 1 & 0 & 0 & 0 & 1 \\ 1 & 1 & 0 & 0 & 1 \\ 1 & 0 & 1 & 0 & 1 \\ 1 & 0 & 0 & 1 & 1 \end{pmatrix}.
$$

In this case, β_g shows the differences between the two groups. The model assumes that the differences are the same over time. The model may be inadequate in RCTs in Sect. 1.1 and Fig. 1.1a, because the distributions of baseline at $j = 1$ are expected to be the same between the groups but the distributions at $j > 1$ will differ.

When the interaction between time and the group is added because the differences between groups are not constant over time, the model becomes

$$Y_{ij} = \beta_0 + \beta_1 x_{1j} + \cdots + \beta_{J-1} x_{J-1j} + \left(\beta_{g0} + \beta_{g1} x_{1j} + \cdots + \beta_{gJ-1} x_{J-1j}\right) x_{gi} + b_i + \varepsilon_{ij}. \tag{1.3.4}$$

In the case of four time points, $\mathbf{X}_i$ for groups A and B and $\boldsymbol{\beta}$ in the model are

$$\mathbf{X}_i = \begin{pmatrix} 1 & 0 & 0 & 0 & 0 & 0 & 0 & 0 \\ 1 & 1 & 0 & 0 & 0 & 0 & 0 & 0 \\ 1 & 0 & 1 & 0 & 0 & 0 & 0 & 0 \\ 1 & 0 & 0 & 1 & 0 & 0 & 0 & 0 \end{pmatrix} \text{ and } \mathbf{X}_i = \begin{pmatrix} 1 & 0 & 0 & 0 & 1 & 0 & 0 & 0 \\ 1 & 1 & 0 & 0 & 1 & 1 & 0 & 0 \\ 1 & 0 & 1 & 0 & 1 & 0 & 1 & 0 \\ 1 & 0 & 0 & 1 & 1 & 0 & 0 & 1 \end{pmatrix},$$

$$\boldsymbol{\beta} = \left(\beta_0, \beta_1, \beta_2, \beta_3, \beta_{g0}, \beta_{g1}, \beta_{g2}, \beta_{g3}\right)^T.$$

In this model, time courses are not assumed to be parallel between the groups.

The expected values at the last time point for groups A and B are

$$E\left(Y_{i4} | x_{gi} = 0\right) = \beta_0 + \beta_3,$$
$$E\left(Y_{i4} | x_{gi} = 1\right) = \beta_0 + \beta_3 + \beta_{g0} + \beta_{g3}.$$

The expected difference at the last time point between groups A and B is

$$E\left(Y_{i4} | x_{gi} = 1\right) - E\left(Y_{i4} | x_{gi} = 0\right) = \beta_{g0} + \beta_{g3}.$$

To estimate the difference, we use the following contrast vector, $\mathbf{L}$,

$$\mathbf{L} = \left(0\ 0\ 0\ 0\ 1\ 0\ 0\ 1\right). \tag{1.3.5}$$

Then, $\beta_{g0} + \beta_{g3}$ is

$$\mathbf{L}\boldsymbol{\beta} = \left(0\ 0\ 0\ 0\ 1\ 0\ 0\ 1\right)\left(\beta_0, \beta_1, \beta_2, \beta_3, \beta_{g0}, \beta_{g1}, \beta_{g2}, \beta_{g3}\right)^T$$
$$= \beta_{g0} + \beta_{g3}. \tag{1.3.6}$$

In Sect. 1.5.5, we discuss the estimation and test using contrasts.

Because the distributions of baseline at $j = 1$ are expected to be the same, we can omit β_{g0} from (1.3.4) in RCTs. The model is

$$Y_{ij} = \beta_0 + \beta_1 x_{1j} + \cdots + \beta_{J-1} x_{J-1j} + \left(\beta_{g1} x_{1j} + \cdots + \beta_{gJ-1} x_{J-1j}\right) x_{gi} + b_i + \varepsilon_{ij}. \tag{1.3.7}$$

In the case of four time points, $\mathbf{X}_i$ for groups A and B and $\boldsymbol{\beta}$ are

$$\mathbf{X}_i = \begin{pmatrix} 1\,0\,0\,0\,0\,0\,0 \\ 1\,1\,0\,0\,0\,0\,0 \\ 1\,0\,1\,0\,0\,0\,0 \\ 1\,0\,0\,1\,0\,0\,0 \end{pmatrix} \text{ and } \mathbf{X}_i = \begin{pmatrix} 1\,0\,0\,0\,0\,0\,0 \\ 1\,1\,0\,0\,1\,0\,0 \\ 1\,0\,1\,0\,0\,1\,0 \\ 1\,0\,0\,1\,0\,0\,1 \end{pmatrix},$$

$$\boldsymbol{\beta} = \left(\beta_0, \beta_1, \beta_2, \beta_3, \beta_{g1}, \beta_{g2}, \beta_{g3} \right)^T.$$

The difference at the last time point between groups A and B is β_{g3}.

1.3.3 Means at Each Time Point with Unstructured Variance Covariance

When there are missing data, the correct specification of the model, including not only the mean structure but also the variance covariance structure, is important, as described in more detail in Sect. 3.2. In the previous sections, we introduced means at each time point with a random intercept. However, the variance covariance structure expressed by the random intercept assumes a constant variance and a constant covariance. These assumptions are too restrictive. Means at each time point with an unstructured (UN) variance covariance are often used recently, because this model has no constraints on the parameters of the mean structure and variance and covariance components. The model is

$$\begin{cases} Y_{ij} = \mu_j + \varepsilon_{ij} \\ \boldsymbol{\varepsilon}_i \sim \mathrm{MVN}(\mathbf{0}, \mathbf{R}_{\mathrm{UN}\,i}) \end{cases}. \tag{1.3.8}$$

Here, μ_j is the mean at the jth $(j = 1, \cdots, J)$ time point, and $\mathbf{R}_{\mathrm{UN}\,i}$ is the UN for $\mathbf{R}_i$. It is assumed that a random error vector, $\boldsymbol{\varepsilon}_i$, follows a multivariate normal distribution with the mean zero and the UN, $\mathbf{R}_{\mathrm{UN}\,i}$.

In the case of four time points, the model for the response, $\mathbf{Y}_i$, and the variance covariance matrix of the response vector, $\mathbf{V}_i = \mathrm{Var}(\mathbf{Y}_i)$, are

$$\mathbf{Y}_i = \begin{pmatrix} Y_{i1} \\ Y_{i2} \\ Y_{i3} \\ Y_{i4} \end{pmatrix} = \begin{pmatrix} 1\,0\,0\,0 \\ 0\,1\,0\,0 \\ 0\,0\,1\,0 \\ 0\,0\,0\,1 \end{pmatrix} \begin{pmatrix} \mu_1 \\ \mu_2 \\ \mu_3 \\ \mu_4 \end{pmatrix} + \begin{pmatrix} \varepsilon_{i1} \\ \varepsilon_{i2} \\ \varepsilon_{i3} \\ \varepsilon_{i4} \end{pmatrix},$$

$$\mathbf{V}_i = \mathbf{R}_i = \begin{pmatrix} \sigma_1^2 & \sigma_{12} & \sigma_{13} & \sigma_{14} \\ \sigma_{12} & \sigma_2^2 & \sigma_{23} & \sigma_{24} \\ \sigma_{13} & \sigma_{23} & \sigma_3^2 & \sigma_{34} \\ \sigma_{14} & \sigma_{24} & \sigma_{34} & \sigma_4^2 \end{pmatrix}.$$

The UN is not parsimonious, and the number of parameters increases largely when the number of time points is large.

For the two group comparison, the models corresponding to (1.3.4) and (1.3.7) are

$$\begin{cases} Y_{ij} = \beta_0 + \beta_1 x_{1j} + \cdots + \beta_{J-1} x_{J-1j} + \left(\beta_{g0} + \beta_{g1} x_{1j} + \cdots + \beta_{gJ-1} x_{J-1j}\right) x_{gi} + \varepsilon_{ij} \\ \varepsilon_i \sim \text{MVN}(\mathbf{0}, \mathbf{R}_{\text{UN}i}) \end{cases},$$

$$(1.3.9)$$

$$\begin{cases} Y_{ij} = \beta_0 + \beta_1 x_{1j} + \cdots + \beta_{J-1} x_{J-1j} + \left(\beta_{g1} x_{1j} + \cdots + \beta_{gJ-1} x_{J-1j}\right) x_{gi} + \varepsilon_{ij} \\ \varepsilon_i \sim \text{MVN}(\mathbf{0}, \mathbf{R}_{\text{UN}i}) \end{cases}.$$

$$(1.3.10)$$

We can assume different UN for each group, but the number of parameters is doubled.

1.3.4 Linear Time Trend Models with Random Intercept and Random Slope

Linear time trend models are also often assumed for the mean structure and random effects. For this assumption, changes per unit time are constant. Let Y_{ij} be the responses corresponding to the jth measurement of the ith ($i = 1, \cdots, N$) subject. t_{ij} is time as a continuous variable. The model is

$$\begin{cases} Y_{ij} = (\beta_0 + b_{0i}) + (\beta_1 + b_{1i})t_{ij} + \varepsilon_{ij} \\ \begin{pmatrix} b_{0i} \\ b_{1i} \end{pmatrix} \sim \text{MVN}\left(\begin{pmatrix} 0 \\ 0 \end{pmatrix}, \begin{pmatrix} \sigma_{G0}^2 & \sigma_{G01} \\ \sigma_{G01} & \sigma_{G1}^2 \end{pmatrix} \right). \\ \varepsilon_{ij} \sim \text{N}(0, \sigma_\varepsilon^2) \end{cases}$$

$$(1.3.11)$$

Here, b_{0i} and b_{1i} are random effects for the intercept and slope for the ith subject, and are called a random intercept and a random slope, respectively. The random effects, $\mathbf{b}_i = (b_{0i}, b_{1i})^T$, are assumed to follow a bivariate normal distribution with the mean vector $\mathbf{0}$, variances σ_{G0}^2 and σ_{G1}^2, and covariance σ_{G01}. The random intercept and slope may be assumed to be independent, that is, $\sigma_{G01} = 0$. The variance of a random error, ε_{ij}, is σ_ε^2, and random errors are assumed to be mutually independent and normally distributed.

In the case of four time points, the model and the variance covariance matrix of the response vector are

$$\mathbf{Y}_i = \begin{pmatrix} Y_{i1} \\ Y_{i2} \\ Y_{i3} \\ Y_{i4} \end{pmatrix} = \begin{pmatrix} 1 & t_{i1} \\ 1 & t_{i2} \\ 1 & t_{i3} \\ 1 & t_{i4} \end{pmatrix} \begin{pmatrix} \beta_0 \\ \beta_1 \end{pmatrix} + \begin{pmatrix} 1 & t_{i1} \\ 1 & t_{i2} \\ 1 & t_{i3} \\ 1 & t_{i4} \end{pmatrix} \begin{pmatrix} b_{i0} \\ b_{i1} \end{pmatrix} + \begin{pmatrix} \varepsilon_{i1} \\ \varepsilon_{i2} \\ \varepsilon_{i3} \\ \varepsilon_{i4} \end{pmatrix},$$

$$\mathbf{V}_i = \mathbf{Z}_i \mathbf{G} \mathbf{Z}_i^T + \mathbf{R}_i$$

$$= \begin{pmatrix} 1 & t_{i1} \\ 1 & t_{i2} \\ 1 & t_{i3} \\ 1 & t_{i4} \end{pmatrix} \begin{pmatrix} \sigma_{G0}^2 & \sigma_{G01} \\ \sigma_{G01} & \sigma_{G1}^2 \end{pmatrix} \begin{pmatrix} 1 & 1 & 1 & 1 \\ t_{i1} & t_{i2} & t_{i3} & t_{i4} \end{pmatrix} + \sigma_\varepsilon^2 \begin{pmatrix} 1 & 0 & 0 & 0 \\ 0 & 1 & 0 & 0 \\ 0 & 0 & 1 & 0 \\ 0 & 0 & 0 & 1 \end{pmatrix}.$$

The diagonal element at the jth time in $\mathbf{V}_i$, that represents the variance, is

$$\sigma_{G0}^2 + 2\sigma_{G01} t_{ij} + \sigma_{G1}^2 t_{ij}^2 + \sigma_\varepsilon^2.$$

The non-diagonal element at the j, kth time in $\mathbf{V}_i$, that represents the covariance, is

$$\sigma_{G0}^2 + \sigma_{G01}\left(t_{ij} + t_{ik}\right) + \sigma_{G1}^2 t_{ij} t_{ik}.$$

Based on these equations, when covariate values of random effects are different across time points, $t_{ij} \neq t_{ik}$ for $j \neq k$, the variance is also different across time points, $\text{Var}(Y_{ij}) \neq \text{Var}(Y_{ik})$, unless $\sigma_{G1}^2 = 0$ and $\sigma_{G01} = 0$. The covariance also depends on the time.

When the time t_{ij} is the same across subjects, $t_{ij} = t_j$, the variance covariance structure is the same across subjects and it is included in the UN. When the time t_{ij} differ across subjects, the variance covariance structure differs across subjects and it is not included in the UN.

Figure 1.3a shows a linear time trend model with a random intercept. We assume a fixed slope ($\sigma_{G1}^2 = 0$, $b_{i1} = 0$). Intercepts are different across subjects but the slopes are the same. Figure 1.3b shows a linear time trend model with a random intercept and a random slope. The variance in response increases with time. In some cases, the variance decreases with time within the observation period.

This model shows a linear time trend and is called a growth curve model. Models with a quadratic equation of time or higher order equations are also called growth curve models. These curves show nonlinear changes according to time but are linear in parameters. In contrast, nonlinear curves such as Gompertz curves and logistic curves show nonlinear changes according to time and are nonlinear in parameters. We discuss nonlinear growth curves further in Chap. 5. Furthermore, growth curves for child development, such as height, weight, and body mass index (BMI), are estimated by other methods.

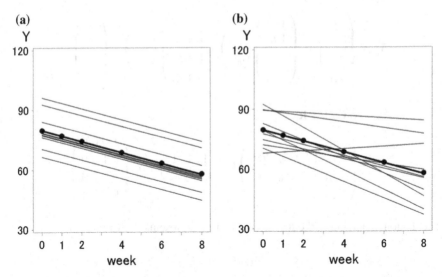

Fig. 1.3 a Linear time trend model with a random intercept. **b** Linear time trend model with a random intercept and a random slope. Estimated means (thick line with closed circles) and predicted values for each subject (thin lines)

1.3.5   Group Comparison Based on Linear Time Trend Models with Random Intercept and Random Slope

This section shows group comparisons based on the linear time trend model (1.3.11). Consider models in which the covariates include time as a continuous variable, the group as a qualitative variable, and an interaction of time and the group. When there are two groups, groups A and B, let x_{gi} be an indicator variable for the group, with $x_{gi} = 0$ in group A and $x_{gi} = 1$ in group B. The linear mixed effects model with the interaction is

$$
\begin{cases}
Y_{ij} = \left(\beta_0 + \beta_{g0}x_{gi} + b_{0i}\right) + \left(\beta_1 + \beta_{g1}x_{gi} + b_{1i}\right)t_{ij} + \varepsilon_{ij} \\[2mm]
\begin{pmatrix} b_{0i} \\ b_{1i} \end{pmatrix} \sim \text{MVN}\left(\begin{pmatrix} 0 \\ 0 \end{pmatrix}, \begin{pmatrix} \sigma_{G0}^2 & \sigma_{G01} \\ \sigma_{G01} & \sigma_{G1}^2 \end{pmatrix} \right) \\[4mm]
\varepsilon_{ij} \sim \text{N}\left(0, \sigma_\varepsilon^2\right)
\end{cases}
\qquad (1.3.12)
$$

For group A, the intercept is β_0 and the slope is β_1. For group B, the intercept is $\beta_0 + \beta_{g0}$, the slope is $\beta_1 + \beta_{g1}$. The coefficient of $x_{gi}t_{ij}$, β_{g1}, is the interaction term between time and the group, and shows the difference in the slopes between the groups. In the case of four time points, the model is

$$
\begin{pmatrix} Y_{i1} \\ Y_{i2} \\ Y_{i3} \\ Y_{i4} \end{pmatrix} = \begin{pmatrix} 1 & x_{gi} & t_{i1} & x_{gi}t_{i1} \\ 1 & x_{gi} & t_{i2} & x_{gi}t_{i2} \\ 1 & x_{gi} & t_{i3} & x_{gi}t_{i3} \\ 1 & x_{gi} & t_{i4} & x_{gi}t_{i4} \end{pmatrix} \begin{pmatrix} \beta_0 \\ \beta_{g0} \\ \beta_1 \\ \beta_{g1} \end{pmatrix} + \begin{pmatrix} 1 & t_{i1} \\ 1 & t_{i2} \\ 1 & t_{i3} \\ 1 & t_{i4} \end{pmatrix} \begin{pmatrix} b_{i0} \\ b_{i1} \end{pmatrix} + \begin{pmatrix} \varepsilon_{i1} \\ \varepsilon_{i2} \\ \varepsilon_{i3} \\ \varepsilon_{i4} \end{pmatrix}.
$$

The design matrices for the fixed effects of groups A and B are

$$
\mathbf{X}_i = \begin{pmatrix} 1 & 0 & t_{i1} & 0 \\ 1 & 0 & t_{i2} & 0 \\ 1 & 0 & t_{i3} & 0 \\ 1 & 0 & t_{i4} & 0 \end{pmatrix} \text{ and } \mathbf{X}_i = \begin{pmatrix} 1 & 1 & t_{i1} & t_{i1} \\ 1 & 1 & t_{i2} & t_{i2} \\ 1 & 1 & t_{i3} & t_{i3} \\ 1 & 1 & t_{i4} & t_{i4} \end{pmatrix}.
$$

Here, the variance covariance parameters, σ_{G0}^2, σ_{G1}^2, σ_{G01}, and σ_ε^2 are assumed to be the same between groups. These can also be assumed to be different between groups.

The slope is often considered a summary measure and is used for group comparisons. Methods to estimate and test the difference in slopes, β_{g1}, have been intensively studied focusing on missing data and dropouts since the late 1980s. In contrast, the asymptote or equilibrium is used in autoregressive linear mixed effects models, and these are potentially useful interpretable summary measures. Funatogawa et al. (2008) studied the estimation of asymptotes focusing on missing data.

1.4 Mean Structures and Variance Covariance Structures

Section 1.3 provides several specific linear mixed effects models that are widely used. This section discusses details of mean structures and variance covariance structures including variable transformation.

1.4.1 Mean Structures

Let μ_{ij} be the mean at the jth time of the ith subject. The mean structure of the model with means at each time point as shown in Sect. 1.3.1 is $\mu_{ij} = \mu_j$, and the number of parameters increases with the number of time points. In the linear time trend model shown in Sect. 1.3.4, $\mu_{ij} = \beta_0 + \beta_1 t_{ij}$, the number of parameters is two: the intercept and slope. A quadratic time trend is

$$
\mu_{ij} = \beta_0 + \beta_1 t_{ij} + \beta_2 t_{ij}^2.
$$

A quadratic equation has a maximum or minimum value, $\beta_0 - \beta_1^2/(2\beta_2)$, at the time, $t_{ij} = -\beta_1/2\beta_2$. After this point, the response changes from an increase to a decrease, or from a decrease to an increase, so that changes are not monotonic. However, sometimes this is not a reasonable assumption, and the fit is not good for later data points. Not only linear or quadratic equations of time but also higher order polynomials of time are used. The lth order polynomial is

$$\mu_{ij} = \sum_{k=0}^{l} \beta_k t_{ij}^k = \beta_0 + \beta_1 t_{ij} + \beta_2 t_{ij}^2 + \cdots + \beta_l t_{ij}^l.$$

A piecewise linear function has linear trends and slope changes at some breakpoints.

In regression analysis, after transformation of a response variable or an explanatory variable, linear models may provide a good fit. Log transformation, $y = \log y$, is often used if the response variable follows a log normal distribution; a logarithm of the response follows a normal distribution. In the area of pharmacokinetics, drug concentrations follow right heavy-tailed distributions, and log transformation is often used. When the variance depends on the mean, log transformation is used in order to stabilize the variance.

The Box–Cox transformation is

$$y^{(\lambda)} = \frac{y^\lambda - 1}{\lambda}, \ (\lambda \neq 0),$$

$$y^{(\lambda)} = \log y, \ (\lambda = 0).$$

Here, the following formula holds:

$$\lim_{\lambda \to 0} \frac{y^\lambda - 1}{\lambda} = \log y.$$

In some cases, frameworks other than linear mixed effects models are more suitable. Autoregressive linear mixed effects models are discussed in the following chapters. Nonlinear mixed effects models are discussed in Chap. 5. Nonparametric regression analysis and smoothing methods are also used.

1.4.2 Variance Covariance Structures

This section discusses variance covariance structures in detail. Table 1.1 shows several variance covariance structures in the case of four time points. The variance covariance matrix of the response vector $\mathbf{Y}_i$ is $\mathbf{V}_i = \mathbf{Z}_i \mathbf{G} \mathbf{Z}_i^T + \mathbf{R}_i$ where $\mathbf{Z}_i \mathbf{G} \mathbf{Z}_i^T$ and $\mathbf{R}_i$ are induced by random effects and random errors, respectively. Variance covariance structures induced by a random intercept are shown in Sect. 1.3.1. The unstructured (UN) is shown in Sect. 1.3.3. Variance covariance structures induced by a random intercept and a random slope are shown in Sect. 1.3.4. The variance covariance structure induced by a random intercept and a random slope with $\begin{pmatrix} t_1 & t_2 & t_3 & t_4 \end{pmatrix} = \begin{pmatrix} 0 & 1 & 2 & 3 \end{pmatrix}$ is given in Table 1.1o. Structures popularly used are independent equal variances, UN, compound symmetry (CS), and first-order autoregressive (AR(1)).

The structure with independent equal variances is used together with random effects. Since data measured from the same subject are usually correlated, the independent structure is not used alone but used together with random effects that account for correlations within a subject. When used with random effects, it is called conditional independence given random effects. The independent structure with unequal variances is also used.

The UN structure has no constraints on variance or covariance parameters. Although this assumption is not strict, the UN is not parsimonious and inefficient. The number of the parameters is large as the number of time points is large. When the number of time points is n_i, the number of parameters is $n_i(n_i + 1)/2$. When the time points increase from n_i to $n_i + 1$, the number of parameters increases by $n_i + 1$.

The CS is also called exchangeable. This structure has two parameters, variance σ^2 and covariance σ_1, which are the same across time points. With other parameters, variance and correlation ρ are the same across time points, respectively. The correlation $\rho = \sigma_1/\sigma^2$ is called the intra-class correlation coefficient. The CS includes the structure induced by a random intercept and independent random errors as shown in Sect. 1.3.1. The diagonal elements are the sum of the between-subject variance and the within-subject variance, and non-diagonal elements are the between-subject variance. Although this structure constrains the covariance and correlation to be positive, these of the CS can be negative. It is a narrower structure compared with CS. The heterogeneous CS (CSH) structure has a CS correlation structure and heterogeneous variances across time points.

The AR(1) structure has a typical serial correlation where the correlation decreases as the time interval increases. The heterogeneous AR(1) (ARH(1)) structure has an AR(1) correlation and heterogeneous variances across time points. The AR(1) structure for $\mathbf{R}_i$, $\mathbf{R}_{\text{AR}i}$, is used in three ways: random errors alone as $\mathbf{V}_i = \mathbf{R}_{\text{AR}i}$, with random effects as $\mathbf{V}_i = \mathbf{Z}_i\mathbf{G}\mathbf{Z}_i^T + \mathbf{R}_{\text{AR}i}$, or with random effects and independent errors as $\mathbf{V}_i = \mathbf{Z}_i\mathbf{G}\mathbf{Z}_i^T + \mathbf{R}_{\text{AR}i} + \sigma^2\mathbf{I}_{n_i}$. There are several approaches to use random effects, serial correlations, and independent errors simultaneously (Diggle 1988; Heitjan 1991; Jones 1993; Funatogawa et al. 2007; Funatogawa et al. 2008). Jones (1993) used serial correlations for continuous time.

The Toeplitz structure has homogeneous variances across time points and homogeneous covariances for equal time distance. The heterogeneous Toeplitz structure has a Toeplitz correlation structure and heterogeneous variances across time points. The j, kth element is $\sigma_j\sigma_k\rho_{|j-k|}$. The two-band Toeplitz structure constrains covariances to be 0 if the time differs over two points. The similar structure is used in autoregressive linear mixed effects models as shown in Sect. 2.4.1 and Table 2.3a to take measurement errors into account.

Kenward (1987) used the first-order ante-dependence (ANTE(1)). The j, kth element is

Table 1.1 Variance covariance structures for four time points

(a) Independent equal variances	(b) Independent unequal variances
$\sigma^2 \begin{pmatrix} 1 & 0 & 0 & 0 \\ 0 & 1 & 0 & 0 \\ 0 & 0 & 1 & 0 \\ 0 & 0 & 0 & 1 \end{pmatrix}$	$\begin{pmatrix} \sigma_1^2 & 0 & 0 & 0 \\ 0 & \sigma_2^2 & 0 & 0 \\ 0 & 0 & \sigma_3^2 & 0 \\ 0 & 0 & 0 & \sigma_4^2 \end{pmatrix}$
(c) Unstructured: UN	(d) Random intercept
$\begin{pmatrix} \sigma_1^2 & \sigma_{12} & \sigma_{13} & \sigma_{14} \\ \sigma_{12} & \sigma_2^2 & \sigma_{23} & \sigma_{24} \\ \sigma_{13} & \sigma_{23} & \sigma_3^2 & \sigma_{34} \\ \sigma_{14} & \sigma_{24} & \sigma_{34} & \sigma_4^2 \end{pmatrix}$	$\begin{pmatrix} \sigma_G^2 & \sigma_G^2 & \sigma_G^2 & \sigma_G^2 \\ \sigma_G^2 & \sigma_G^2 & \sigma_G^2 & \sigma_G^2 \\ \sigma_G^2 & \sigma_G^2 & \sigma_G^2 & \sigma_G^2 \\ \sigma_G^2 & \sigma_G^2 & \sigma_G^2 & \sigma_G^2 \end{pmatrix}$
(e) Random intercept and independent equal variances, inter- and intra-variances	(f) Compound symmetry: CS, variance and covariance
$\begin{pmatrix} \sigma_G^2+\sigma_\varepsilon^2 & \sigma_G^2 & \sigma_G^2 & \sigma_G^2 \\ \sigma_G^2 & \sigma_G^2+\sigma_\varepsilon^2 & \sigma_G^2 & \sigma_G^2 \\ \sigma_G^2 & \sigma_G^2 & \sigma_G^2+\sigma_\varepsilon^2 & \sigma_G^2 \\ \sigma_G^2 & \sigma_G^2 & \sigma_G^2 & \sigma_G^2+\sigma_\varepsilon^2 \end{pmatrix}$	$\begin{pmatrix} \sigma^2 & \sigma_1 & \sigma_1 & \sigma_1 \\ \sigma_1 & \sigma^2 & \sigma_1 & \sigma_1 \\ \sigma_1 & \sigma_1 & \sigma^2 & \sigma_1 \\ \sigma_1 & \sigma_1 & \sigma_1 & \sigma^2 \end{pmatrix}$
(g) CS, variance and correlation	(h) Heterogeneous CS: CSH
$\sigma^2 \begin{pmatrix} 1 & \rho & \rho & \rho \\ \rho & 1 & \rho & \rho \\ \rho & \rho & 1 & \rho \\ \rho & \rho & \rho & 1 \end{pmatrix}$	$\begin{pmatrix} \sigma_1^2 & \sigma_1\sigma_2\rho & \sigma_1\sigma_3\rho & \sigma_1\sigma_4\rho \\ \sigma_1\sigma_2\rho & \sigma_2^2 & \sigma_2\sigma_3\rho & \sigma_2\sigma_4\rho \\ \sigma_1\sigma_3\rho & \sigma_2\sigma_3\rho & \sigma_3^2 & \sigma_3\sigma_4\rho \\ \sigma_1\sigma_4\rho & \sigma_2\sigma_4\rho & \sigma_3\sigma_4\rho & \sigma_4^2 \end{pmatrix}$
(i) First-order autoregressive: AR(1)	(j) Heterogeneous AR(1): ARH(1)
$\sigma^2 \begin{pmatrix} 1 & \rho & \rho^2 & \rho^3 \\ \rho & 1 & \rho & \rho^2 \\ \rho^2 & \rho & 1 & \rho \\ \rho^3 & \rho^2 & \rho & 1 \end{pmatrix}$	$\begin{pmatrix} \sigma_1^2 & \sigma_1\sigma_2\rho & \sigma_1\sigma_3\rho^2 & \sigma_1\sigma_4\rho^3 \\ \sigma_1\sigma_2\rho & \sigma_2^2 & \sigma_2\sigma_3\rho & \sigma_2\sigma_4\rho^2 \\ \sigma_1\sigma_3\rho^2 & \sigma_2\sigma_3\rho & \sigma_3^2 & \sigma_3\sigma_4\rho \\ \sigma_1\sigma_4\rho^3 & \sigma_2\sigma_4\rho^2 & \sigma_3\sigma_4\rho & \sigma_4^2 \end{pmatrix}$
(k) Toeplitz	(l) Heterogeneous Toeplitz
$\begin{pmatrix} \sigma^2 & \sigma_1 & \sigma_2 & \sigma_3 \\ \sigma_1 & \sigma^2 & \sigma_1 & \sigma_2 \\ \sigma_2 & \sigma_1 & \sigma^2 & \sigma_1 \\ \sigma_3 & \sigma_2 & \sigma_1 & \sigma^2 \end{pmatrix}$	$\begin{pmatrix} \sigma_1^2 & \sigma_1\sigma_2\rho_1 & \sigma_1\sigma_3\rho_2 & \sigma_1\sigma_4\rho_3 \\ \sigma_1\sigma_2\rho_1 & \sigma_2^2 & \sigma_2\sigma_3\rho_1 & \sigma_2\sigma_4\rho_2 \\ \sigma_1\sigma_3\rho_2 & \sigma_2\sigma_3\rho_1 & \sigma_3^2 & \sigma_3\sigma_4\rho_1 \\ \sigma_1\sigma_4\rho_3 & \sigma_2\sigma_4\rho_2 & \sigma_3\sigma_4\rho_1 & \sigma_4^2 \end{pmatrix}$

(continued)

Table 1.1   (continued)

(m) Two-band Toeplitz	(n) First-order ante-dependence: ANTE(1)[a]

$$\begin{pmatrix} \sigma^2 & \sigma_1 & 0 & 0 \\ \sigma_1 & \sigma^2 & \sigma_1 & 0 \\ 0 & \sigma_1 & \sigma^2 & \sigma_1 \\ 0 & 0 & \sigma_1 & \sigma^2 \end{pmatrix}$$

$$\begin{pmatrix} \sigma_1^2 & \sigma_1\sigma_2\rho_1 & \sigma_1\sigma_3\rho_1\rho_2 & \sigma_1\sigma_4\rho_1\rho_2\rho_3 \\ & \sigma_2^2 & \sigma_2\sigma_3\rho_2 & \sigma_2\sigma_4\rho_2\rho_3 \\ & & \sigma_3^2 & \sigma_3\sigma_4\rho_3 \\ & & & \sigma_4^2 \end{pmatrix}$$

(o) Random intercept and random slope with $\left(t_1 \ t_2 \ t_3 \ t_4 \right) = \left(0 \ 1 \ 2 \ 3 \right)$[a]

$$\begin{pmatrix} \sigma_{G0}^2 & \sigma_{G0}^2 + \sigma_{G01} & \sigma_{G0}^2 + 2\sigma_{G01} & \sigma_{G0}^2 + 3\sigma_{G01} \\ & \sigma_{G0}^2 + 2\sigma_{G01} + \sigma_{G1}^2 & \sigma_{G0}^2 + 3\sigma_{G01} + 2\sigma_{G1}^2 & \sigma_{G0}^2 + 4\sigma_{G01} + 3\sigma_{G1}^2 \\ & & \sigma_{G0}^2 + 4\sigma_{G01} + 4\sigma_{G1}^2 & \sigma_{G0}^2 + 5\sigma_{G01} + 6\sigma_{G1}^2 \\ & & & \sigma_{G0}^2 + 6\sigma_{G01} + 9\sigma_{G1}^2 \end{pmatrix}$$

[a]Lower triangular elements are omitted

$$\sigma_j\sigma_k \prod_{l=j}^{k-1} \rho_l.$$

When the number of time points is n_i, the number of parameters is $2n_i - 1$. The ANTE(1) is non-stationary, where the variances are not constant across time points and the correlations depend on time points. In contrast, the AR(1), CS, and Toeplitz are stationary, where the variances are constant across time points and the correlations depend on only time distance.

Variance covariance structures using a stochastic process are applied. In the Ornstein–Uhlenbeck process (OU process), the variance and covariance between $Y(s)$ and $Y(t)$ are

$$\mathrm{Cov}(Y(s), Y(t)) = \sigma^2(2\alpha)^{-1}e^{-\alpha|t-s|}.$$

This continuous time process corresponds to AR(1) for discrete time with $\rho = e^{-\alpha}$. The following process that integrates the OU process is called the integrated Ornstein–Uhlenbeck process (IOU process),

$$W(t) = \int_0^t Y(u)du.$$

The variance of $W(t)$ and the covariance between $W(s)$ and $W(t)$ are

$$\mathrm{Var}(W(t)) = \sigma^2\alpha^{-3}\left(\alpha t + e^{-\alpha t} - 1\right),$$

$$\text{Cov}(W(s), W(t)) = \sigma^2 (2\alpha^3)^{-1} \{2\alpha \min(s, t) + e^{-\alpha t} + e^{-\alpha s} - 1 - e^{-\alpha|t-s|}\}.$$

Taylor et al. (1994) and Taylor and Law (1998) used the IOU process and Sy et al. (1997) used the bivariate extension.

Variance covariance matrices are sometimes assumed to be different across levels of a factor such as a group. If there are two groups, the number of parameters increases by two times. Results of statistical tests or estimation may largely depend on this assumption. The unequal variances in the analysis of covariance are discussed in Funatogawa et al. (2011) and Funatogawa and Funatogawa (2011).

1.5 Inference

1.5.1 Maximum Likelihood Method

For the estimation of linear mixed effects models, maximum likelihood (ML) methods are often used. For simplicity, this section first explains likelihood for independent data. Let Y follow a normal distribution with the mean μ and variance σ^2. The probability density function of Y is

$$f(Y) = \frac{1}{\sqrt{2\pi\sigma^2}} \exp\left\{-\frac{1}{2}\left(\frac{Y - \mu}{\sigma}\right)^2\right\}. \tag{1.5.1}$$

The probability density function is a function of a random variable Y given the parameters (μ, σ^2). It shows what value of Y tends to occur. Now, let $Y_1, \cdots, Y_N$ be N random variables that follow independently an identical normal distribution with the mean μ and variance σ^2. Then, the probability density function of $Y_1, \cdots, Y_N$ is

$$f(Y_1, \cdots, Y_N) = \prod_{i=1}^{N} \frac{1}{\sqrt{2\pi\sigma^2}} \exp\left\{-\frac{1}{2}\left(\frac{Y_i - \mu}{\sigma}\right)^2\right\}. \tag{1.5.2}$$

Because of the independent variables, this is a simple multiplication of the above probability density function. The likelihood function is algebraically the same as the probability density function, but a function of parameters (μ, σ^2) given the data, $Y_1, \cdots, Y_N$. The ML method maximizes the log-likelihood (ll) for the estimation of unknown parameters. The log-likelihood is

$$ll_{\text{ML}} = -\frac{N}{2}\log(2\pi) - \frac{N}{2}\log\sigma^2 - \frac{1}{2}\sum_{i=1}^{N}\left(\frac{Y_i - \mu}{\sigma}\right)^2. \tag{1.5.3}$$

Minus two times log-likelihood ($-2ll$) is used for the calculation,

$$-2ll_{\text{ML}} = N \log(2\pi) + N \log \sigma^2 + \sum_{i=1}^{N} \left(\frac{Y_i - \mu}{\sigma} \right)^2. \qquad (1.5.4)$$

The ML estimator (MLE) of μ is the arithmetic mean, $\bar{Y} = \sum_{i=1}^{N} Y_i/N$. The MLE of σ^2 is $\sum_{i=1}^{N}(Y_i - \bar{Y})^2/N$, and is biased because it does not take account of the decrease of one degree of freedom by the estimation of the mean parameter. It is known that the unbiased estimator of σ^2 is $\sum_{i=1}^{N}(Y_i - \bar{Y})^2/(N-1)$. When N is infinite, both converge to the same value. In this simple example, the MLE of σ^2 is biased but consistent.

Next, the likelihood for longitudinal data is provided. In longitudinal data analysis, data from different subjects are often assumed to be independent, but data from the same subject are not assumed to be independent. In linear mixed effects models, the marginal distribution of $\mathbf{Y}_i$ is a multivariate normal distribution with the mean $\mathbf{X}_i\boldsymbol{\beta}$ and variance covariance matrix $\mathbf{V}_i = \mathbf{Z}_i\mathbf{G}\mathbf{Z}_i^T + \mathbf{R}_i$. Then, the probability density function of $\mathbf{Y}_1, \cdots, \mathbf{Y}_N$ is

$$f(\mathbf{Y}_1, \cdots, \mathbf{Y}_N) = \prod_{i=1}^{N} (2\pi)^{-\frac{1}{2}} |\mathbf{V}_i|^{-\frac{1}{2}} \exp\left\{ -\frac{1}{2}(\mathbf{Y}_i - \mathbf{X}_i\boldsymbol{\beta})^T \mathbf{V}_i^{-1} (\mathbf{Y}_i - \mathbf{X}_i\boldsymbol{\beta}) \right\},$$

$$(1.5.5)$$

where $|\mathbf{V}_i|$ is the determinant of $\mathbf{V}_i$. The marginal log-likelihood function, ll_{ML}, and $-2ll_{\text{ML}}$ are

$$ll_{\text{ML}} = -\frac{\sum_{i=1}^{N} n_i}{2} \log(2\pi) - \frac{1}{2} \sum_{i=1}^{N} \log|\mathbf{V}_i| - \frac{1}{2} \sum_{i=1}^{N} (\mathbf{Y}_i - \mathbf{X}_i\boldsymbol{\beta})^T \mathbf{V}_i^{-1} (\mathbf{Y}_i - \mathbf{X}_i\boldsymbol{\beta}),$$

$$(1.5.6)$$

$$-2ll_{\text{ML}} = \sum_{i=1}^{N} \left\{ n_i \log(2\pi) + \log|\mathbf{V}_i| + (\mathbf{Y}_i - \mathbf{X}_i\boldsymbol{\beta})^T \mathbf{V}_i^{-1} (\mathbf{Y}_i - \mathbf{X}_i\boldsymbol{\beta}) \right\}. \qquad (1.5.7)$$

The ML estimates are obtained from maximizing ll_{ML} or minimizing $-2ll_{\text{ML}}$ with respect to unknown parameters. When variance covariance parameters are known, the MLEs of the fixed effects parameters $\boldsymbol{\beta}$ obtained from minimizing $-2ll_{\text{ML}}$ are

$$\hat{\boldsymbol{\beta}} = \left(\sum_{i=1}^{N} \mathbf{X}_i^T \mathbf{V}_i^{-1} \mathbf{X}_i \right)^{-} \sum_{i=1}^{N} \mathbf{X}_i^T \mathbf{V}_i^{-1} \mathbf{Y}_i, \qquad (1.5.8)$$

where the superscript "$-$" means a generalized inverse. The variance covariance parameters are usually unknown. To obtain the ML estimates of the variance covariance parameters, $\boldsymbol{\beta}$ is concentrated out of the likelihood by substituting the equation of $\hat{\boldsymbol{\beta}}$ (1.5.8) in $-2ll_{\text{ML}}$ (1.5.7). Then, we minimize the following concentrated $-2ll$:

$$-2ll_{\text{ML CONC}} = \sum_{i=1}^{N} \left\{ n_i \log(2\pi) + \log|\mathbf{V}_i| + \mathbf{r}_i^T \mathbf{V}_i^{-1} \mathbf{r}_i \right\}, \qquad (1.5.9)$$

where $\mathbf{r}_i = \mathbf{Y}_i - \mathbf{X}_i \hat{\boldsymbol{\beta}}$. We can reduce the number of unknown parameters by p. p is the number of fixed effects parameters. Variance covariance parameters are not usually solved explicitly and iteration methods such as the Newton–Raphson method, the expectation-maximization (EM) algorithm, or the Fisher's scoring algorithm are used. The MLEs of the fixed effects parameters $\boldsymbol{\beta}$ are

$$\hat{\boldsymbol{\beta}} = \left(\sum_{i=1}^{N} \mathbf{X}_i^T \hat{\mathbf{V}}_i^{-1} \mathbf{X}_i \right)^{-} \sum_{i=1}^{N} \mathbf{X}_i^T \hat{\mathbf{V}}_i^{-1} \mathbf{Y}_i, \qquad (1.5.10)$$

where $\mathbf{V}_i$ in Eq. (1.5.8) is replaced with the ML estimates $\hat{\mathbf{V}}_i$.

If $\mathbf{V}_i$ can be written as $\sigma^2 \mathbf{V}_{ci}$, σ^2 can be concentrated out of the likelihood. σ^2 has a closed form given $\boldsymbol{\beta}$ and $\mathbf{V}_{ci}$, and it is substituted in $-2ll_{\text{ML}}$. $\mathbf{V}_i^{-1}$ in (1.5.8) is replaced by $\mathbf{V}_{ci}^{-1}$, and $\hat{\boldsymbol{\beta}}$ is substituted in $-2ll_{\text{ML}}$. The unknown parameters in $-2ll_{\text{ML CONC}}$ is $\mathbf{V}_{ci}^{-1}$. This reduces further the number of optimization parameters by one. The MLEs of the fixed effects parameters are

$$\hat{\boldsymbol{\beta}} = \left(\sum_{i=1}^{N} \mathbf{X}_i^T \hat{\mathbf{V}}_{ci}^{-1} \mathbf{X}_i \right)^{-} \sum_{i=1}^{N} \mathbf{X}_i^T \hat{\mathbf{V}}_{ci}^{-1} \mathbf{Y}_i. \qquad (1.5.11)$$

An example in the state space form is given in Sect. 6.5.2.

The ML estimates of variance covariance components are biased because these do not take account of the decrease of degree of freedoms by the estimation of fixed effects parameters. Therefore, the restricted maximum likelihood (REML) method is used. The log-likelihood function for REML is

$$ll_{\text{REML}} = -\frac{\sum_{i=1}^{N} n_i - p}{2} \log(2\pi) - \frac{1}{2} \sum_{i=1}^{N} \log|\mathbf{V}_i| - \frac{1}{2} \log \left| \sum_{i=1}^{N} \mathbf{X}_i^T \mathbf{V}_i^{-1} \mathbf{X}_i \right|$$

$$- \frac{1}{2} \sum_{i=1}^{N} (\mathbf{Y}_i - \mathbf{X}_i \hat{\boldsymbol{\beta}})^T \mathbf{V}_i^{-1} (\mathbf{Y}_i - \mathbf{X}_i \hat{\boldsymbol{\beta}}). \qquad (1.5.12)$$

The REML method estimates the variance covariance parameters based on the residual contrast. The residual contrast is a linear combination of $\mathbf{Y}_i$ that does not depend on $\boldsymbol{\beta}$. It is also called the residual maximum likelihood (REML) method. The residual contrast is shown in Sect. 1.6 using vector representations. The REML method cannot be used to compare goodness of fit for two models with different fixed effects because log-likelihoods of the two models use different residual contrasts.

1.5.2 Variances of Estimates of Fixed Effects

When variance covariance parameters are known, the variance covariance matrix of the ML estimates of fixed effects vector is

$$\text{Var}(\hat{\boldsymbol{\beta}}) = \left(\sum_{i=1}^{N} \mathbf{X}_i^T \mathbf{V}_i^{-1} \mathbf{X}_i \right)^{-} \left(\sum_{i=1}^{N} \mathbf{X}_i^T \mathbf{V}_i^{-1} \text{Var}(\mathbf{Y}_i) \mathbf{V}_i^{-1} \mathbf{X}_i \right) \left(\sum_{i=1}^{N} \mathbf{X}_i^T \mathbf{V}_i^{-1} \mathbf{X}_i \right)^{-}.$$

$$(1.5.13)$$

If $\mathbf{V}_i = \text{Var}(\mathbf{Y}_i)$,

$$\text{Var}(\hat{\boldsymbol{\beta}}) = \left(\sum_{i=1}^{N} \mathbf{X}_i^T \mathbf{V}_i^{-1} \mathbf{X}_i \right)^{-}. \qquad (1.5.14)$$

The variance and covariance are replaced by the ML estimates $\hat{\mathbf{V}}_i$,

$$\text{Var}(\hat{\boldsymbol{\beta}}) = \left(\sum_{i=1}^{N} \mathbf{X}_i^T \hat{\mathbf{V}}_i^{-1} \mathbf{X}_i \right)^{-}. \qquad (1.5.15)$$

Since this variance covariance matrix is based on likelihood, the mean structure, the variance covariance structure, and the distribution assumption in the linear mixed effects model need to be correct. The standard errors of $\hat{\boldsymbol{\beta}}$ based on this equation are underestimated because this equation does not take account of the uncertainty in the estimation of the variance and covariance.

Even if $\mathbf{V}_i = \text{Var}(\mathbf{Y}_i)$ is wrongly specified, the following sandwich estimator provides a consistent estimator of $\text{Var}(\hat{\boldsymbol{\beta}})$:

$$\text{Var}(\hat{\boldsymbol{\beta}}) = \left(\sum_{i=1}^{N} \mathbf{X}_i^T \hat{\mathbf{V}}_i^{-1} \mathbf{X}_i \right)^{-} \left\{ \sum_{i=1}^{N} \mathbf{X}_i^T \hat{\mathbf{V}}_i^{-1} \left(\mathbf{Y}_i - \mathbf{X}_i \hat{\boldsymbol{\beta}} \right) \left(\mathbf{Y}_i - \mathbf{X}_i \hat{\boldsymbol{\beta}} \right)^T \hat{\mathbf{V}}_i^{-1} \mathbf{X}_i \right\}$$

$$\left(\sum_{i=1}^{N} \mathbf{X}_i^T \hat{\mathbf{V}}_i^{-1} \mathbf{X}_i \right)^{-}. \qquad (1.5.16)$$

It is also called robust variance.

1.5.3 Prediction

The joint distribution of $\mathbf{Y}_i$ and $\mathbf{b}_i$ is the following multivariate normal distribution:

$$\begin{pmatrix} \mathbf{Y}_i \\ \mathbf{b}_i \end{pmatrix} \sim \text{MVN} \left(\begin{pmatrix} \mathbf{X}_i \boldsymbol{\beta} \\ \mathbf{0} \end{pmatrix}, \begin{pmatrix} \mathbf{Z}_i \mathbf{G} \mathbf{Z}_i^T + \mathbf{R}_i & \mathbf{Z}_i \mathbf{G} \\ \mathbf{G} \mathbf{Z}_i^T & \mathbf{G} \end{pmatrix} \right). \qquad (1.5.17)$$

From this distribution, the conditional expectation of $\mathbf{b}_i$ given $\mathbf{Y}_i$ is

$$E(\mathbf{b}_i|\mathbf{Y}_i) = \mathbf{G}\mathbf{Z}_i^T\mathbf{V}_i^{-1}(\mathbf{Y}_i - \mathbf{X}_i\boldsymbol{\beta}). \qquad (1.5.18)$$

Replacing $\mathbf{V}_i$, $\mathbf{G}$, and $\boldsymbol{\beta}$ by the ML estimates, $\hat{\mathbf{V}}_i$, $\hat{\mathbf{G}}$, and $\hat{\boldsymbol{\beta}}$, the predictors of random effects are

$$\hat{\mathbf{b}}_i = \hat{\mathbf{G}}\mathbf{Z}_i^T\hat{\mathbf{V}}_i^{-1}(\mathbf{Y}_i - \mathbf{X}_i\hat{\boldsymbol{\beta}}). \qquad (1.5.19)$$

When variance covariance parameters are known, $\hat{\boldsymbol{\beta}}$ in Eq. (1.5.8) is the best linear unbiased estimator (BLUE) and $E(\mathbf{b}_i|\mathbf{Y}_i)$ is the best linear unbiased predictor (BLUP). Since $\mathbf{b}_i$ is a random vector, it is called a predictor but not an estimator. The term "best" means the minimum error variance among linear unbiased estimators or predictors. Since variance covariance parameters are unknown, these are replaced by ML or REML estimates. $\hat{\boldsymbol{\beta}}$ in Eq. (1.5.10) is called empirical BLUE (EBLUE) and $\hat{\mathbf{b}}_i$ is called empirical BLUP (EBLUP). These are based on empirical Bayes methods.

The response profile $\hat{\mathbf{Y}}_i$ in the ith subject is predicted by

$$\begin{aligned}
\hat{\mathbf{Y}}_i &= \mathbf{X}_i\hat{\boldsymbol{\beta}} + \mathbf{Z}_i\hat{\mathbf{b}}_i \\
&= (\hat{\mathbf{R}}_i\hat{\mathbf{V}}_i^{-1})\mathbf{X}_i\hat{\boldsymbol{\beta}} + (\mathbf{I}_{n_i} - \hat{\mathbf{R}}_i\hat{\mathbf{V}}_i^{-1})\mathbf{Y}_i.
\end{aligned} \qquad (1.5.20)$$

This is a weighted mean of the population mean $\mathbf{X}_i\hat{\boldsymbol{\beta}}$ and observed response $\mathbf{Y}_i$. It is called shrinkage because the predicted values shrink to the population mean. The extent of the shrinkage depends on the relative size of $\mathbf{R}_i$ and $\mathbf{V}_i = \mathbf{Z}_i\mathbf{G}\mathbf{Z}_i^T + \mathbf{R}_i$. The larger intra-subject variance compared with the inter-subject variance results in the larger weight on $\mathbf{X}_i\hat{\boldsymbol{\beta}}$. The larger inter-subject variance results in the larger weight on $\mathbf{Y}_i$. The larger number of observations n_i in the ith subject results in the smaller shrinkage.

Similarly, the predicted values of the random effects $\hat{\mathbf{b}}_i$ in the ith subject are a weighted mean of the REML estimates of the fixed effects parameter $\hat{\boldsymbol{\beta}}$ and the ordinary least square estimates of the corresponding parameters $\hat{\boldsymbol{\beta}}_{\text{OLS}i}$ based on only the data of the ith subject. When $\mathbf{X}_i = \mathbf{Z}_i$ and $\mathbf{R}_i = \sigma^2\mathbf{I}_{n_i}$, the estimator of $\boldsymbol{\beta}_i = \boldsymbol{\beta} + \mathbf{b}_i$ is

$$\begin{aligned}
\hat{\boldsymbol{\beta}}_i &= \hat{\boldsymbol{\beta}} + \hat{\mathbf{b}}_i \\
&= \mathbf{W}_i\hat{\boldsymbol{\beta}}_{\text{OLS}i} + (\mathbf{I}_q - \mathbf{W}_i)\hat{\boldsymbol{\beta}},
\end{aligned} \qquad (1.5.21)$$

where

$$\mathbf{W}_i = \mathbf{G}\left\{\mathbf{G} + \sigma^2(\mathbf{Z}_i^T\mathbf{Z}_i)^{-1}\right\}^{-1}. \qquad (1.5.22)$$

The larger inter-subject variance compared with the intra-subject variance results in the closer estimates of $\hat{\boldsymbol{\beta}}_i$ to $\hat{\boldsymbol{\beta}}_{\text{OLS}i}$. The smaller inter-subject variance results in the closer estimates of $\hat{\mathbf{b}}_i$ to $\mathbf{0}$.

1.5.4 Goodness of Fit for Models

There are several indicators for the goodness of fit. Akaike's information criterion (AIC) and Schwartz's Bayesian information criterion (BIC) are measures of goodness of fit that put a penalty on an increase in the number of parameters. Let ll_{MLmax} and $ll_{REMLmax}$ be the maximum value of the log-likelihood for ML and REML, $K = \sum_{i=1}^{N} n_i$ be the number of data, p be the number of parameters for fixed effects, and q be the number of parameters for random effects. For ML and REML, AIC and BIC are

$$\text{AIC}_{ML} = -2ll_{MLmax} + 2(p+q), \qquad (1.5.23)$$

$$\text{AIC}_{REML} = -2ll_{REMLmax} + 2q, \qquad (1.5.24)$$

$$\text{BIC}_{ML} = -2ll_{MLmax} + (p+q)\log K, \qquad (1.5.25)$$

$$\text{BIC}_{REML} = -2ll_{REMLmax} + q\log(K-p). \qquad (1.5.26)$$

The REML method cannot be used to compare two models with different fixed effects as described in Sect. 1.5.1.

1.5.5 Estimation and Test Using Contrast

Let $\mathbf{L}$ be a $1 \times q$ contrast vector and consider the estimation of $\mathbf{L}\boldsymbol{\beta}$. $\mathbf{L}\boldsymbol{\beta}$ is assumed to be estimable such that $\mathbf{L}\boldsymbol{\beta} = \mathbf{k}E(\mathbf{Y}) = \mathbf{k}\mathbf{X}\boldsymbol{\beta}$ for some vector of constants, $\mathbf{k}$, where $\mathbf{Y} = \left(\mathbf{Y}_1^T, \cdots, \mathbf{Y}_N^T\right)^T$ and $\mathbf{X} = \left(\mathbf{X}_1^T, \cdots, \mathbf{X}_N^T\right)^T$. The estimator is $\mathbf{L}\hat{\boldsymbol{\beta}}$, and the two-sided 95% confidence interval is

$$\mathbf{L}\hat{\boldsymbol{\beta}} \pm t_{\nu(0.975)} \sqrt{\mathbf{L}\left(\sum_{i=1}^{N} \mathbf{X}_i^T \hat{\mathbf{V}}_i^{-1} \mathbf{X}_i\right)^{-} \mathbf{L}^T}, \qquad (1.5.27)$$

where $t_{\nu(0.975)}$ is the upper 97.5th percentile of the t distribution with ν degrees of freedom (df). Using the contrast vector $\mathbf{L}$, a t test with the null hypothesis of $\mathbf{L}\boldsymbol{\beta} = 0$ can be performed. The following test statistic approximately follows a t distribution with ν degrees of freedom,

$$\frac{\mathbf{L}\hat{\boldsymbol{\beta}}}{\sqrt{\mathbf{L}\left(\sum_{i=1}^{N} \mathbf{X}_i^T \hat{\mathbf{V}}_i^{-1} \mathbf{X}_i\right)^{-} \mathbf{L}^T}}. \qquad (1.5.28)$$

The degree of freedom, ν, usually needs to be estimated by an approximation. There are several approximation methods such as the Satterthwaite approximation (Satterthwaite 1946). The Kenward–Roger method (Kenward and Roger 1997) uses an adjusted estimator of the variance covariance matrix to reduce small sample bias.

When there are multiple contrasts using a $k \times p$ $(p \geq k)$ full rank matrix $\mathbf{L}$, an F test with the null hypothesis of $\mathbf{L}\boldsymbol{\beta} = \mathbf{0}$ can be performed. The following test statistic approximately follows an F distribution:

$$\frac{\hat{\boldsymbol{\beta}}^T \mathbf{L}^T \left\{ \mathbf{L} \left(\sum_{i=1}^N \mathbf{X}_i^T \hat{\mathbf{V}}_i^{-1} \mathbf{X}_i \right)^{-} \mathbf{L}^T \right\}^{-} \mathbf{L}\hat{\boldsymbol{\beta}}}{\text{rank}(\mathbf{L})}. \tag{1.5.29}$$

The numerator degree of freedom is the rank of $\mathbf{L}$, rank($\mathbf{L}$). The denominator degree of freedom usually needs to be estimated using an approximation. When $k = 1$, the test statistic of the F test is the square of the test statistic of the t test.

1.6   Vector Representation

In the previous sections, linear mixed effects models are shown using the vector $\mathbf{Y}_i$ for each subject. This section shows the representation using the vector $\mathbf{Y} = \left(\mathbf{Y}_1^T, \cdots, \mathbf{Y}_N^T \right)^T$. Let $\mathbf{X} = \left(\mathbf{X}_1^T, \cdots, \mathbf{X}_N^T \right)^T$, $\mathbf{b} = \left(\mathbf{b}_1^T, \cdots, \mathbf{b}_N^T \right)^T$, and $\boldsymbol{\varepsilon} = \left(\boldsymbol{\varepsilon}_1^T, \cdots, \boldsymbol{\varepsilon}_N^T \right)^T$, $\mathbf{Z} = \text{diag}(\mathbf{Z}_i)$. The linear mixed effects models shown in Sect. 1.2 are expressed by

$$\mathbf{Y} = \mathbf{X}\boldsymbol{\beta} + \mathbf{Z}\mathbf{b} + \boldsymbol{\varepsilon}, \tag{1.6.1}$$

$$\begin{pmatrix} \mathbf{Y}_1 \\ \mathbf{Y}_2 \\ \vdots \\ \mathbf{Y}_N \end{pmatrix} = \begin{pmatrix} \mathbf{X}_1 \\ \mathbf{X}_2 \\ \vdots \\ \mathbf{X}_N \end{pmatrix} \begin{pmatrix} \beta_1 \\ \beta_2 \\ \vdots \\ \beta_p \end{pmatrix} + \begin{pmatrix} \mathbf{Z}_1 & 0 & \cdots & 0 \\ 0 & \mathbf{Z}_2 & & 0 \\ \vdots & & \ddots & \\ 0 & 0 & & \mathbf{Z}_N \end{pmatrix} \begin{pmatrix} \mathbf{b}_1 \\ \mathbf{b}_2 \\ \vdots \\ \mathbf{b}_N \end{pmatrix} + \begin{pmatrix} \boldsymbol{\varepsilon}_1 \\ \boldsymbol{\varepsilon}_2 \\ \vdots \\ \boldsymbol{\varepsilon}_N \end{pmatrix}. \tag{1.6.2}$$

The variance covariance matrices, $\mathbf{V} = \text{Var}(\mathbf{Y})$, $\mathbf{G}_A = \text{Var}(\mathbf{b})$, and $\mathbf{R} = \text{Var}(\boldsymbol{\varepsilon})$, are

$$\mathbf{V} = \text{diag}(\mathbf{V}_i) = \begin{pmatrix} \mathbf{V}_1 & 0 & \cdots & 0 \\ 0 & \mathbf{V}_2 & & 0 \\ \vdots & & \ddots & \\ 0 & 0 & & \mathbf{V}_N \end{pmatrix}, \tag{1.6.3}$$

$$\mathbf{G}_A = \text{diag}(\mathbf{G}_i) = \begin{pmatrix} \mathbf{G}_1 & 0 & \cdots & 0 \\ 0 & \mathbf{G}_2 & & 0 \\ \vdots & & \ddots & \\ 0 & 0 & & \mathbf{G}_N \end{pmatrix} = \begin{pmatrix} \mathbf{G} & 0 & \cdots & 0 \\ 0 & \mathbf{G} & & 0 \\ \vdots & & \ddots & \\ 0 & 0 & & \mathbf{G} \end{pmatrix}, \tag{1.6.4}$$

$$\mathbf{R} = \mathrm{diag}(\mathbf{R}_i) = \begin{pmatrix} \mathbf{R}_1 & 0 & \cdots & 0 \\ 0 & \mathbf{R}_2 & & 0 \\ \vdots & & \ddots & \\ 0 & 0 & & \mathbf{R}_N \end{pmatrix}, \tag{1.6.5}$$

where $\mathbf{G}_i = \mathbf{G}$ and $\mathbf{V} = \mathbf{Z}\mathbf{G}_A\mathbf{Z}^T + \mathbf{R}$.
$-2ll_{\mathrm{ML}}$, $-2ll_{\mathrm{REML}}$, $\hat{\boldsymbol{\beta}}$, $\mathrm{Var}(\hat{\boldsymbol{\beta}})$, and $\hat{\mathbf{b}}$ shown in Sect. 1.5 are expressed by

$$-2ll_{\mathrm{ML}} = \left(\sum_{i=1}^{N} n_i\right) \log(2\pi) + \log|\mathbf{V}| + (\mathbf{Y} - \mathbf{X}\boldsymbol{\beta})^T \mathbf{V}^{-1}(\mathbf{Y} - \mathbf{X}\boldsymbol{\beta}), \tag{1.6.6}$$

$$-2ll_{\mathrm{REML}} = \left(\sum_{i=1}^{N} n_i - p\right) \log(2\pi) + \log|\mathbf{V}| + \log\left|\mathbf{X}^T\mathbf{V}^{-1}\mathbf{X}\right|$$

$$+ \left(\mathbf{Y} - \mathbf{X}\hat{\boldsymbol{\beta}}\right)^T \mathbf{V}^{-1}\left(\mathbf{Y} - \mathbf{X}\hat{\boldsymbol{\beta}}\right), \tag{1.6.7}$$

$$\hat{\boldsymbol{\beta}} = \left(\mathbf{X}^T\mathbf{V}^{-1}\mathbf{X}\right)^{-}\mathbf{X}^T\mathbf{V}^{-1}\mathbf{Y}, \tag{1.6.8}$$

$$\mathrm{Var}(\hat{\boldsymbol{\beta}}) = \left(\mathbf{X}^T\mathbf{V}^{-1}\mathbf{X}\right)^{-}, \tag{1.6.9}$$

$$\hat{\mathbf{b}} = \hat{\mathbf{G}}_A\mathbf{Z}^T\hat{\mathbf{V}}^{-1}\left(\mathbf{Y} - \mathbf{X}\hat{\boldsymbol{\beta}}\right). \tag{1.6.10}$$

$\hat{\boldsymbol{\beta}}$ and $\hat{\mathbf{b}}_i$ as shown in Sect. 1.5 can be also derived from the following mixed model equation:

$$\begin{pmatrix} \mathbf{X}^T\hat{\mathbf{R}}^{-1}\mathbf{X} & \mathbf{X}^T\hat{\mathbf{R}}^{-1}\mathbf{Z} \\ \mathbf{Z}^T\hat{\mathbf{R}}^{-1}\mathbf{X} & \mathbf{Z}^T\hat{\mathbf{R}}^{-1}\mathbf{Z} + \hat{\mathbf{G}}_A^{-1} \end{pmatrix}\begin{pmatrix} \hat{\boldsymbol{\beta}} \\ \hat{\mathbf{b}} \end{pmatrix} = \begin{pmatrix} \mathbf{X}^T\hat{\mathbf{R}}^{-1}\mathbf{Y} \\ \mathbf{Z}^T\hat{\mathbf{R}}^{-1}\mathbf{Y} \end{pmatrix}. \tag{1.6.11}$$

Now, we explain residual maximum likelihood (REML) in Sect. 1.5.1. Let $\mathbf{K}$ be a $\sum_{i=1}^{N} n_i \times \left(\sum_{i=1}^{N} n_i - p\right)$ full rank matrix that satisfies $\mathbf{K}^T\mathbf{X} = \mathbf{0}$. $\mathbf{K}^T\mathbf{Y}$ is called residual contrast. It follows a multivariate normal distribution with the mean vector $\mathbf{0}$ and variance covariance matrix $\mathbf{K}^T\mathbf{V}\mathbf{K}$, and does not depend on $\boldsymbol{\beta}$. The log-likelihood of $\mathbf{K}^T\mathbf{Y}$ is ll_{REML}.

References

Diggle PJ (1988) An approach to the analysis of repeated measurements. Biometrics 44:959–971
Diggle PJ, Heagerty P, Liang KY, Zeger SL (2002) Analysis of longitudinal data, 2nd edn. Oxford University Press
Diggle PJ, Liang KY, Zeger SL (1994) Analysis of longitudinal data. Oxford University Press
Dwyer JH, Feinleib M, Lippert P, Hoffmeister H (eds) (1992) Statistical models for longitudinal studies of health. Oxford University Press

Fitzmaurice GM, Davidian M, Verbeke G, Molenberghs G (eds) (2009) Longitudinal data analysis. Chapman & Hall/CRC Press

Fitzmaurice GM, Laird NM, Ware JH (2004) Applied longitudinal analysis. Wiley

Fitzmaurice GM, Laird NM, Ware JH (2011) Applied longitudinal analysis, 2nd edn. Wiley

Funatogawa I, Funatogawa T (2011) Analysis of covariance with pre-treatment measurements in randomized trials: comparison of equal and unequal slopes. Biometrical J 53:810–821

Funatogawa I, Funatogawa T, Ohashi Y (2007) An autoregressive linear mixed effects model for the analysis of longitudinal data which show profiles approaching asymptotes. Stat Med 26:2113–2130

Funatogawa T, Funatogawa I, Shyr Y (2011) Analysis of covariance with pre-treatment measurements in randomized trials under the cases that covariances and post-treatment variances differ between groups. Biometrical J 53:512–524

Funatogawa T, Funatogawa I, Takeuchi M (2008) An autoregressive linear mixed effects model for the analysis of longitudinal data which include dropouts and show profiles approaching asymptotes. Stat Med 27:6351–6366

Gregoire TG, Brillinger DR, Diggle PJ, Russek-Cohen E, Warren WG, Wolfinger RD (eds) (1997) Modelling longitudinal and spatially correlated data. Springer-Verlag

Hand D, Crowder M (1996) Practical longitudinal data analysis. Chapman & Hall

Heitjan DF (1991) Nonlinear modeling of serial immunologic data: a case study. J Am Stat Assoc 86:891–898

Jones RH (1993) Longitudinal data with serial correlation: a state-space approach. Chapman & Hall

Kenward MG (1987) A method for comparing profiles of repeated measurements. Appl Stat 36:296–308

Kenward MG, Roger JH (1997) Small sample inference for fixed effects from restricted maximum likelihood. Biometrics 53:983–997

Laird NM (2004) Analysis of longitudinal & cluster-correlated data. IMS

Laird NM, Ware JH (1982) Random-effects models for longitudinal data. Biometrics 38:963–974

Littell RC, Miliken GA, Stroup WW, Wolfinger RD (1996) SAS system for mixed models. SAS Institute Inc

Littell RC, Miliken GA, Stroup WW, Wolfinger RD, Schabenberger O (2006) SAS for mixed models, 2nd edn. SAS Institute Inc

Satterthwaite FE (1946) An approximate distribution of estimates of variance components. Biometrics Bull 2:110–114

Sy JP, Taylor JMG, Cumberland WG (1997) A stochastic model for analysis of bivariate longitudinal AIDS data. Biometrics 53:542–555

Tango T (2017) Repeated measures design with generalized linear mixed models for randomized controlled trials. CRC Press

Taylor JMG, Cumberland WG, Sy JP (1994) A stochastic model for analysis of longitudinal AIDS data. J Am Stat Assoc 89:727–736

Taylor JMG, Law N (1998) Does the covariance structure matter in longitudinal modeling for the prediction of future CD4 counts? Stat Med 17:2381–2394

Verbeke G, Molenberghs G (eds) (1997) Linear mixed models in practice—a SAS oriented approach. Springer-Verlag

Verbeke G, Molenberghs G (2000) Linear mixed models for longitudinal data. Springer-Verlag

Vonesh EF (2012) Generalized linear and nonlinear models for correlated data. Theory and applications using SAS. SAS Institute Inc

Wu H, Zhang J-T (2006) Nonparametric regression methods for longitudinal data analysis. Mixed-effects modeling approaches. Wiley

Zimmerman DL, Núñez-Antón VA (2010) Antedependence models for longitudinal data. CRC Press

Chapter 2
Autoregressive Linear Mixed Effects Models

Abstract In the previous chapter, longitudinal data analysis using linear mixed effects models was discussed. This chapter discusses autoregressive linear mixed effects models in which the current response is regressed on the previous response, fixed effects, and random effects. These are an extension of linear mixed effects models and autoregressive models. Autoregressive models regressed on the response variable itself have two remarkable properties: approaching asymptotes and state-dependence. Asymptotes can be modeled by fixed effects and random effects. The current response depends on current covariates and past covariate history. Three vector representations of autoregressive linear mixed effects models are provided: an autoregressive form, response changes with asymptotes, and a marginal form which is unconditional on previous responses. The marginal interpretation is the same with subject specific interpretation as well as linear mixed effects models. Variance covariance structures corresponding to AR(1) errors, measurement errors, and random effects in the baseline and asymptote are presented. Likelihood of marginal and autoregressive forms for maximum likelihood estimation are also provided. The marginal form can be used even if there are intermittent missing values. We discuss the difference between autoregressive models of the response itself which focused in this book and models with autoregressive error terms.

Keywords Asymptote · Autoregressive
Autoregressive linear mixed effects model · Longitudinal · State-dependence

2.1 Autoregressive Models of Response Itself

2.1.1 Introduction

There are three major approaches for modeling longitudinal data: mixed effects models, marginal models, and transition models (Diggle et al. 2002; Fitzmaurice et al. 2011). Linear mixed effects models and marginal models with linear mean structures are discussed in the framework of linear mixed effects models through

Chap. 1. We discuss nonlinear mixed effects models in Chap. 5. Mixed effects models include both fixed effects and random effects, and random effects take account for variability across subjects. Marginal models directly model the marginal distribution of the response without random effects. The model with the unstructured variance covariance in Sect. 1.3.3 is an example of marginal models. In the case of a linear model, interpretation of the fixed effects parameter is the same between two models. Although linear mixed effects models have good properties, they are not always satisfactory to express nonlinear time trends. On the other hand, nonlinear mixed effects models with nonlinear random effects parameters are complicated and there is discrepancy between marginal and subject specific interpretation. Autoregressive linear mixed effects models in this book simply express nonlinear time trends which gradually move toward an asymptote without an approximation and the discrepancy in interpretation. The marginal distribution is explicitly given, and it is useful because more interest is often taken in the marginal profile than the profile conditional on the previous response. The autoregression in the response itself changes a static mixed effects model into a dynamic one. It provides one of the simplest models that take into account the past covariate history.

There are two types of transition (autoregressive) models for continuous response variables: autoregressive models of the response itself and models with autoregressive error terms. Let $Y_{i,t}$, $\mathbf{X}_{i,t}$, and $\varepsilon_{i,t}$ be the response, a design vector, and an error term for the subject i at time t. In autoregressive models of the response itself, the current response, $Y_{i,t}$, is regressed on the previous response, $Y_{i,t-1}$,

$$Y_{i,t} = \rho Y_{i,t-1} + \mathbf{X}_{i,t}\boldsymbol{\beta} + \varepsilon_{i,t}. \tag{2.1.1}$$

In linear models with the first-order autoregressive error, AR(1) error, an error term, $\varepsilon_{ei,t}$, is regressed on the previous error term, $\varepsilon_{ei,t-1}$,

$$\begin{cases} Y_{i,t} = \mathbf{X}_{i,t}\boldsymbol{\beta}_e + \varepsilon_{ei,t} \\ \varepsilon_{ei,t} = \rho\varepsilon_{ei,t-1} + \eta_{i,t} \end{cases}. \tag{2.1.2}$$

As discussed in Sect. 1.4.2, AR(1) errors are used in linear mixed effects models and marginal models, and classified into these models too. For example, the model (2.1.2) belongs to marginal models. The interpretation of the fixed effects parameter $\boldsymbol{\beta}_e$ is the same with linear mixed effects models and marginal models. However, transition models of the response itself provide different interpretations of the fixed effects parameter. In this book, the autoregressive models of the response itself are discussed in detail. The differences from models with AR(1) errors are discussed in Sect. 2.6. If the response is a discrete variable, such as a binary or count variable, the interpretation of the fixed effects parameter differs across the three major approaches.

Here, we describe the difference of notation from Chap. 1. In Chap. 1, $\mathbf{Y}_i = \left(Y_{i1}, Y_{i2}, \cdots, Y_{in_i}\right)^T$ is the $n_i \times 1$ vector of the response corresponding to the ith $(i = 1, \cdots, N)$ subject measured from 1 to n_i occasions, and Y_{ij} is the jth measurement. In autoregressive linear mixed effects models, the baseline is an important concept, and we model the baseline measurement separately from the measurement at later time points. For these models in the following chapters, we define $Y_{i,0}$ is a baseline measurement, and $Y_{i,t}$ ($t = 1, 2, \cdots, T_i$) is the tth measurement after the baseline measurement. $\mathbf{Y}_i = \left(Y_{i,0}, Y_{i,1}, Y_{i,2}, \cdots, Y_{i,T_i}\right)^T$ is the $(T_i + 1) \times 1$ vector of the response.

The regression models that regress the current response on the previous response and covariates have been called by various names: autoregressive models (Rosner et al. 1985; Rosner and Muñoz 1988), conditional models (Rosner and Muñoz 1992), conditional autoregressive models (Schmid 1996), state-dependence models (Lindsey 1993), transition models (Diggle et al. 2002), dynamic models (Anderson and Hsiao 1982; Schmid 2001), Markov models, autoregressive response models, lagged-response models (Rabe-Hesketh and Skrondal 2012), and so on. We shall call them autoregressive models in this book.

Although processes other than AR(1) are not used much in longitudinal data analysis, other processes are used in time series analysis. We introduce related process briefly. An autoregressive moving average process of order (p, q) (ARMA(p, q)) is

$$Y_t = \rho_1 Y_{t-1} + \cdots + \rho_p Y_{t-p} + \xi_t + \theta_1 \xi_{t-1} + \cdots + \theta_q \xi_{t-q},$$

where $Y_t, Y_{t-1}, \cdots, Y_{t-p}$ are observed values, $\xi_t, \xi_{t-1}, \cdots, \xi_{t-q}$ are a random variable with the mean zero and constant variance, $\rho_1, \cdots, \rho_p$ are autoregressive parameters, and $\theta_1, \cdots, \theta_q$ are also unknown parameters. The AR(1) is ARMA(1,0). An autoregressive process of order p (AR(p)) is ARMA$(p, 0)$ and

$$Y_t = \rho_1 Y_{t-1} + \cdots + \rho_p Y_{t-p} + \xi_t.$$

A moving average process of order q (MA(q)) is ARMA$(0, q)$ and

$$Y_t = \xi_t + \theta_1 \xi_{t-1} + \cdots + \theta_q \xi_{t-q}.$$

In this book, we use only the AR(1) process.

Section 2.1 discusses autoregressive models for one subject. Section 2.2 introduces random effects to account for variability across subjects in longitudinal data. Section 2.3 introduces autoregressive linear mixed effects models with vector representations which make clear many aspects of the models including the relationship with linear mixed effects models. Section 2.4 provides details of variance covariance structures. Section 2.5 provides estimation methods. Section 2.6 discusses models with autoregressive error terms.

2.1.2 Response Changes in Autoregressive Models

This section shows how the response level changes in an autoregressive model and the interpretation of parameters. For simplicity, we consider a case where there is only one subject and neither random effect nor random error. First, we show an autoregressive model with an intercept. The response at time t $(t = 1, 2, \cdots, T)$, Y_t, is a linear function of the previous response, Y_{t-1}, as

$$Y_t = \rho Y_{t-1} + \beta_{\text{int}}, \qquad (2.1.3)$$

where ρ is a regression coefficient of the previous response and β_{int} is an intercept. These are unknown parameters.

Assuming $\rho \neq 1$, this model can be transformed as

$$Y_t - Y_{t-1} = (1 - \rho)\left(\frac{\beta_{\text{int}}}{1 - \rho} - Y_{t-1}\right). \qquad (2.1.4)$$

If Y_{t-1} equals $(1 - \rho)^{-1}\beta_{\text{int}}$, the change is zero, and $(1 - \rho)^{-1}\beta_{\text{int}} \equiv Y_{\text{Asy}}$ can be interpreted as an asymptote if $0 < \rho < 1$. Y_{Asy} is a parameter. The response changes with asymptotes are

$$\begin{cases} Y_t - Y_{t-1} = (1 - \rho)\left(Y_{\text{Asy}} - Y_{t-1}\right) \\ Y_{\text{Asy}} = (1 - \rho)^{-1}\beta_{\text{int}} \end{cases}, \qquad (2.1.5)$$

where $Y_{\text{Asy}} - Y_{t-1}$ is the size remaining to the asymptote. The expected change from Y_{t-1} to Y_t is proportional to the remaining size with a proportional constant $(1 - \rho)$. This interpretation is biologically important, and the model corresponds to monomolecular (Mitscherlich) growth curves in continuous time. The curves are nonlinear in the parameter ρ, and the relationship with nonlinear models is discussed in Sect. 5.1. An asymptote is also called equilibrium or level at a steady state. Y_t is an internally dividing point that divides a segment line $Y_{t-1}Y_{\text{Asy}}$ into $1 - \rho : \rho$, as

$$Y_t = \rho Y_{t-1} + (1 - \rho)Y_{\text{Asy}}. \qquad (2.1.6)$$

Figure 2.1 shows a numerical example. The model is $Y_t = 0.6Y_{t-1} + 2$ with $Y_0 = 1$. The asymptote is $(1 - \rho)^{-1}\beta_{\text{int}} = 5$. $Y_1 = 2.6$, $Y_2 = 3.56$, $Y_3 = 4.136$, and $Y_4 = 4.4816$. As shown in the figure, autoregressive models represent a profile with an initial sharp change, gradually decreasing rates of change, and approaching to asymptote. For describing such phenomena, linear time trend models are not sufficient. Although quadratic or higher order polynomial models sometimes show adequate fit, their parameters are hard to interpret.

The model can also be transformed into the marginal form without the previous response. Given Y_0, Y_t is

$$Y_t = \rho^t Y_0 + \sum_{l=1}^{t} \rho^{t-l}\beta_{\text{int}}, \qquad (2.1.7)$$

where $\sum_{l=1}^{t} \rho^{t-l} = \left(1 - \rho^t\right)/(1 - \rho)$. If Y_0 is modeled separately from Y_t $(t > 0)$ as $Y_0 = \beta_{\text{base}}$, the marginal form is

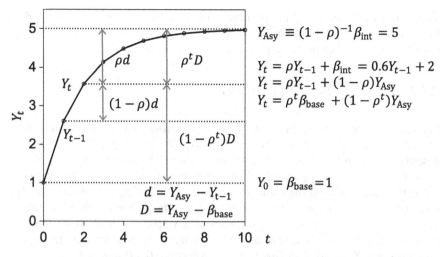

Fig. 2.1 Autoregressive model, $Y_t = \rho Y_{t-1} + \beta_{\text{int}}$ with $Y_0 = \beta_{\text{base}}$. $Y_{\text{Asy}} \equiv (1-\rho)^{-1}\beta_{\text{int}}$ is the asymptote. The change $Y_t - Y_{t-1}$ is proportional to the size remaining to the asymptote $Y_{\text{Asy}} - Y_{t-1}$ with a proportional constant $(1-\rho)$. Y_t is an internally dividing point that divides $Y_{t-1}Y_{\text{Asy}}$ into $1-\rho : \rho$ and $\beta_{\text{base}}Y_{\text{Asy}}$ into $1-\rho^t : \rho^t$

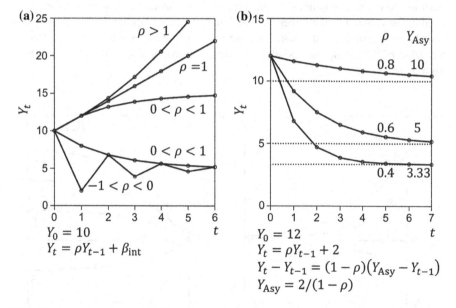

Fig. 2.2 Autoregressive model, $Y_t = \rho Y_{t-1} + \beta_{\text{int}}$. **a** Effects of the autoregressive coefficient ρ. **b** Effects of the autoregressive coefficient ρ $(0 < \rho < 1)$. ρ affects both the asymptote, $(1-\rho)^{-1}\beta_{\text{int}}$, and the proportion of the change, $(1-\rho)$

$$Y_t = \rho^t \beta_{\text{base}} + \left(1 - \rho^t\right)\beta_{\text{int}}/(1-\rho)$$
$$= \rho^t \beta_{\text{base}} + \left(1 - \rho^t\right)Y_{\text{Asy}}. \tag{2.1.8}$$

This equation also shows that the asymptote, Y_{Asy}, can be expressed by $(1 - \rho)^{-1}\beta_{int}$, because $Y_t \to (1 - \rho)^{-1}\beta_{int}$ when $t \to \infty$ if $0 < \rho < 1$. Y_t is an internally dividing point that divides a segment line $\beta_{base}\,Y_{Asy}$ into $1 - \rho^t : \rho^t$.

2.1.3 Interpretation of Parameters

Figure 2.2a shows how changes in the response over time depend on the autoregressive coefficient, ρ, in the model, $Y_t = \rho Y_{t-1} + \beta_{int}$. When $0 < \rho < 1$, as previously described, the response level changes to the asymptote, $Y_{Asy} \equiv (1 - \rho)^{-1}\beta_{int}$. When $\rho = 1$, $Y_t - Y_{t-1} = \beta_{int}$ and the change per unit of time is β_{int}. This shows a linear time trend with an intercept of β_{base} and a slope of β_{int}. When $\rho > 1$, the response does not converge to an asymptote but instead shows a diverging trend. When $-1 < \rho < 0$, the response becomes an asymptote with amplitude. In this book, we consider only the case of $0 < \rho < 1$. When there is no intercept term, β_{int}, the equation is $Y_t = \rho Y_{t-1}$ and the response changes to 0 if $0 < \rho < 1$.

Figure 2.2b shows how changes in the response over time depend on the autoregressive coefficient, ρ, in the model, $Y_t = \rho Y_{t-1} + \beta_{int}$, under the constraint $0 < \rho < 1$. The asymptote is $(1 - \rho)^{-1}\beta_{int}$, and changes in the response are proportional to the size remaining, with a proportional constant $(1 - \rho)$. Therefore, the smaller the value of ρ, the larger the absolute difference between the baseline value and the asymptote, and the faster the approach to the asymptote. Thus, ρ affects the value of the asymptote and also the speed, or proportion, of the change to the asymptote.

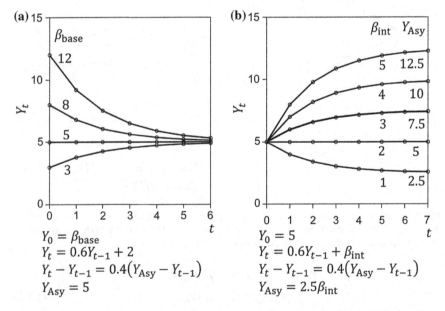

Fig. 2.3 Autoregressive model, $Y_t = \rho Y_{t-1} + \beta_{int}$ with $Y_0 = \beta_{base}$. **a** Effects of the baseline. **b** Effects of the intercept. The asymptote, $(1 - \rho)^{-1}\beta_{int}$, depends on the intercept

In this book, the parameter ρ is a constant value and does not randomly change across subjects. In more general situations, ρ can be a random variable. It should be noted that the asymptotes change simultaneously as ρ changes, and that there is a constraint of $0 < \rho < 1$ in order to interpret the model as a representation of profiles approaching the asymptotes.

Figure 2.3a, b shows how changes in the response depend on the baseline and intercept, respectively, in the model, $Y_t = \rho Y_{t-1} + \beta_{int}$ with $Y_0 = \beta_{base}$. The baseline parameter defines the response at time 0, but the effect is diminishing. Because the number of time points is limited in longitudinal data, the baseline is important. The intercept defines the asymptote $(1 - \rho)^{-1}\beta_{int}$. The proportion of the change to the asymptote is constant, $(1 - \rho)$, but the change itself is larger when the remaining size is larger. The remaining size is

$$Y_{Asy} - Y_{t-1} = \rho^{t-1}\{(1-\rho)^{-1}\beta_{int} - \beta_{base}\}$$
$$= \rho^{t-1}\{Y_{Asy} - \beta_{base}\}. \qquad (2.1.9)$$

Figure 2.4a shows how changes in the response depend on the time-independent covariates in the model, $Y_t = \rho Y_{t-1} + \beta_{int} + \beta_{cov}x$. Asymptotes, $(1-\rho)^{-1}(\beta_{int} + \beta_{cov}x)$, linearly depend on the covariate, x, with the coefficient, $(1-\rho)^{-1}\beta_{cov}$. The proportion of the change to the asymptote is the same, $(1 - \rho)$, but the change is larger if the remaining size is larger. Figure 2.4b shows how response changes when time as a continuous variable is the time-dependent covariate. In this case, the asymptote changes linearly with time.

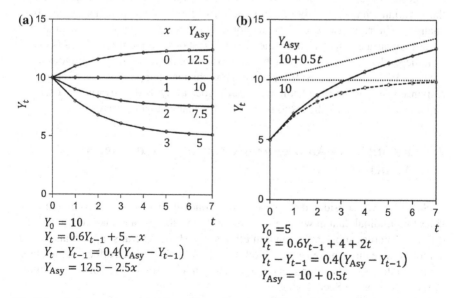

Fig. 2.4 **a** Autoregressive model, $Y_t = \rho Y_{t-1} + \beta_{int} + \beta_{cov}x$. Effects of a time-independent covariate. The asymptote, $(1 - \rho)^{-1}(\beta_{int} + \beta_{cov}x)$, linearly depends on the covariate, x, with the coefficient $(1 - \rho)^{-1}\beta_x$. **b** Autoregressive model, $Y_t = \rho Y_{t-1} + \beta_{int} + \beta_{cov}t$. The time-dependent covariate is time, t, as a continuous variable

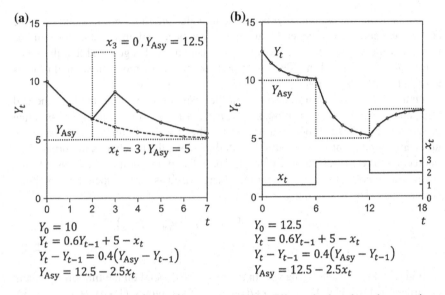

Fig. 2.5 Autoregressive model, $Y_t = \rho Y_{t-1} + \beta_{int} + \beta_{cov} x_t$. Effects of a time-dependent covariate. **a** Two time courses: the covariate changes at one time point and no covariate changes. **b** Response changes to the new asymptote when the covariate value changes

Figure 2.5a, b shows how changes in the response depend on a time-dependent covariate. For example, the response level will change depending on changes in drug doses. Figure 2.5a shows two time courses in the response level: one is under the covariate change at only one time point, while the other is under no covariate changes. The response level depends not only on current covariate values but also on past covariate values through the previous response in the model. This is called state-dependence. Figure 2.5b shows response changes to the new asymptote when the covariate value changes. If the covariate is changed temporally, this effect on the response lasts for the time being and finally goes away. If the covariate is changed and kept at some level, the responses gradually change to a new asymptote.

2.2 Examples of Autoregressive Linear Mixed Effects Models

In Sect. 2.1, we considered the case of only one subject. In this section, we consider longitudinal data in which there are multiple subjects, and take account of the difference across subjects by random effects. We show several simple examples of autoregressive linear mixed effects models. Hereafter, βs are fixed effects parameters, bs are random effects parameters, and εs are random errors. We discuss structures of random errors in Sect. 2.4.1.

2.2.1 *Example Without Covariates*

An example of autoregressive linear mixed effects models without covariates is

$$\begin{cases} Y_{i,0} = \beta_{\text{base}} + b_{\text{base } i} + \varepsilon_{i,0} \\ Y_{i,t} = \rho Y_{i,t-1} + (\beta_{\text{int}} + b_{\text{int } i}) + \varepsilon_{i,t}, \ (t > 0) \end{cases}, \tag{2.2.1}$$

where $b_{\text{base } i}$ and $b_{\text{int } i}$ are random effects showing the differences across subjects, and are assumed to be normally distributed. These equations can be represented by the equation of baseline and that of response changes with the asymptotes,

$$\begin{cases} Y_{i,0} = \beta_{\text{base}} + b_{\text{base } i} + \varepsilon_{i,0} \\ Y_{i,t} - Y_{i,t-1} = (1 - \rho)\left(Y_{\text{Asy } i} - Y_{i,t-1}\right) + \varepsilon_{i,t}, (t > 0). \\ Y_{\text{Asy } i} = (1 - \rho)^{-1}(\beta_{\text{int}} + b_{\text{int } i}) \end{cases} \tag{2.2.2}$$

These equations can also be represented by the marginal form,

$$\begin{cases} Y_{i,0} = \beta_{\text{base}} + b_{\text{base } i} + \varepsilon_{\text{m } i,0} \\ Y_{i,t} = \rho^t(\beta_{\text{base}} + b_{\text{base } i}) + \sum_{l=1}^{t} \rho^{t-l}(\beta_{\text{int}} + b_{\text{int } i}) + \varepsilon_{\text{m } i,t}, (t > 0) \end{cases}, \tag{2.2.3}$$

where $\varepsilon_{\text{m } i,t}$ is

$$\varepsilon_{\text{m } i,t} = \sum_{l=0}^{t} \rho^{t-l} \varepsilon_{i,l}. \tag{2.2.4}$$

In particular, $\varepsilon_{\text{m } i,0} = \varepsilon_{i,0}$. Here, m in subscript means the marginal. The following expression holds:

$$\varepsilon_{\text{m } i,t} = \rho \varepsilon_{\text{m } i,t-1} + \varepsilon_{i,t}. \tag{2.2.5}$$

Figure 2.6 shows examples where error terms are omitted. By including random effects, we can represent the changes from a baseline response level to another response level for each subject. These kinds of changes are seen in studies in which the effects of intervention are examined. The variability across subjects at the baseline and asymptote are $\text{Var}(b_{\text{base } i})$ and $\text{Var}\left((1 - \rho)^{-1}b_{\text{int } i}\right)$, respectively. The mean structures are the same between Fig. 2.6a and b. The variability across subjects is larger at the baseline compared to later time points in Fig. 2.6a, but it is less at the baseline in Fig. 2.6b. The correlation between the baseline and asymptote is large in some cases and small in others. When the effects of an intervention are examined, variability across subjects changes because of the differences in the response to the intervention across subjects.

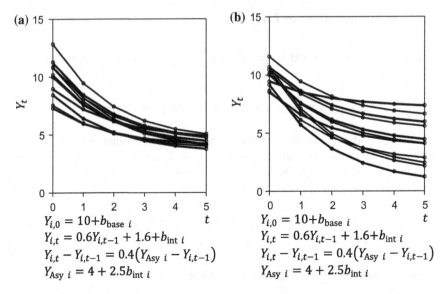

$$Y_{i,0} = 10 + b_{\text{base } i}$$
$$Y_{i,t} = 0.6 Y_{i,t-1} + 1.6 + b_{\text{int } i}$$
$$Y_{i,t} - Y_{i,t-1} = 0.4 \left(Y_{\text{Asy } i} - Y_{i,t-1} \right)$$
$$Y_{\text{Asy } i} = 4 + 2.5 b_{\text{int } i}$$

Fig. 2.6 Autoregressive linear mixed effects model—error terms are omitted, $Y_{i,0} = \beta_{\text{base}} + b_{\text{base } i}$, $Y_{i,t} = \rho Y_{i,t-1} + \beta_{\text{int}} + b_{\text{int } i}$. $b_{\text{base } i}$ is a random baseline and $b_{\text{int } i}$ is a random intercept. **a** Larger variance in a random baseline. **b** Larger variance in a random asymptote, $(1 - \rho)^{-1} b_{\text{int } i}$

2.2.2 Example with Time-Independent Covariates

We provide an example of autoregressive linear mixed effects models with time-independent covariates. We model the baseline and the later time points separately. The value of the time-independent covariate in each subject remains the same over time. In the example, the response changes are compared between groups A and B. The covariates are dummy variables indicating each group. Let $x_{\text{base } i} = 1$ for $t = 0$ and the subject i in group B and 0 otherwise. Let $x_{\text{int } i} = 1$ for $t > 0$ and the subject i in group B and 0 otherwise. The model is

$$\begin{cases} Y_{i,0} = \beta_{\text{base}} + \beta_{\text{base g}} x_{\text{base } i} + b_{\text{base } i} + \varepsilon_{i,0} \\ Y_{i,t} = \rho Y_{i,t-1} + \beta_{\text{int}} + \beta_{\text{int g}} x_{\text{int } i} + b_{\text{int } i} + \varepsilon_{i,t}, \ (t > 0) \end{cases} . \tag{2.2.6}$$

The asymptote of the subject i, $Y_{\text{Asy } i}$, is

$$Y_{\text{Asy } i} = (1 - \rho)^{-1} \left(\beta_{\text{int}} + \beta_{\text{int g}} x_{\text{int } i} + b_{\text{int } i} \right). \tag{2.2.7}$$

The expected values of the asymptotes of groups A and B are

$$(1 - \rho)^{-1} \beta_{\text{int}},$$
$$(1 - \rho)^{-1} \left(\beta_{\text{int}} + \beta_{\text{int g}} \right).$$

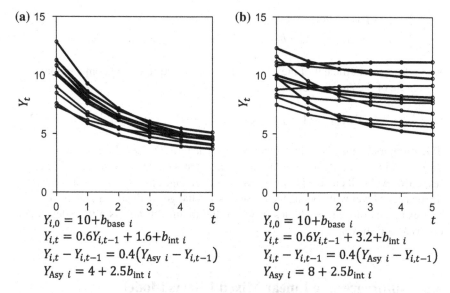

(a)
$$Y_{i,0} = 10 + b_{\text{base } i}$$
$$Y_{i,t} = 0.6 Y_{i,t-1} + 1.6 + b_{\text{int } i}$$
$$Y_{i,t} - Y_{i,t-1} = 0.4 \left(Y_{\text{Asy } i} - Y_{i,t-1} \right)$$
$$Y_{\text{Asy } i} = 4 + 2.5 b_{\text{int } i}$$

(b)
$$Y_{i,0} = 10 + b_{\text{base } i}$$
$$Y_{i,t} = 0.6 Y_{i,t-1} + 3.2 + b_{\text{int } i}$$
$$Y_{i,t} - Y_{i,t-1} = 0.4 \left(Y_{\text{Asy } i} - Y_{i,t-1} \right)$$
$$Y_{\text{Asy } i} = 8 + 2.5 b_{\text{int } i}$$

Fig. 2.7 Autoregressive linear mixed effects model under randomization—error terms are omitted, $Y_{i,0} = \beta_{\text{base}} + b_{\text{base } i}$, $Y_{i,t} = \rho Y_{i,t-1} + \beta_{\text{int}} + \beta_{\text{int g}} x_g + b_{\text{int } i}$. **a, b** The response changes in each subject in two groups with $x_g = 0$ in **a** and $x_g = 1$ in **b**. Baseline distributions are similar between groups but distributions at later times are not. The variances in a random baseline $b_{\text{base } i}$ are similar but the variances in a random asymptote $(1 - \rho)^{-1} b_{\text{int } i}$ are not

The expected difference between the asymptotes of groups A and B is

$$(1 - \rho)^{-1} \beta_{\text{int g}}.$$

An example of the comparison across three treatment groups including a placebo group in a randomized controlled trial (RCT) is shown in Sect. 3.1. In some cases, variance covariance matrices are assumed to differ across treatment groups. For example, Fig. 2.7 shows an illustration of an RCT where error terms are omitted. In RCTs, the distribution of the responses at baseline is expected to be similar between the two groups, but the distribution of the responses at later time points may largely differ. In this case, we can assume $\beta_{\text{base g}} = 0$, the equal variances of $b_{\text{base } i}$, and the different variances of $b_{\text{int } i}$ between the two groups.

2.2.3 Example with a Time-Dependent Covariate

As shown in Fig. 2.5, the asymptotes change according to changes in a time-dependent covariate. An example of a time-dependent covariate is drug dosing in a clinical study. Let $x_{i,t}$ be a drug dose for the subject i at time t. The following equations are an example of autoregressive linear mixed effects models with a time-dependent covariate:

$$\begin{cases} Y_{i,0} = \beta_{\text{base}} + b_{\text{base}\,i} + \varepsilon_{i,0} \\ Y_{i,t} = \rho Y_{i,t-1} + (\beta_{\text{int}} + b_{\text{int}\,i}) + (\beta_{\text{cov}} + b_{\text{cov}\,i})x_{i,t} + \varepsilon_{i,t}, \ (t > 0) \end{cases}, \qquad (2.2.8)$$

where $b_{\text{cov}\,i}$, a coefficient of the covariate $x_{i,t}$, is an additional random variable. The asymptote of the subject i at time t, $Y_{\text{Asy}\,i,t}$, is

$$Y_{\text{Asy}\,i,t} = (1 - \rho)^{-1}\{\beta_{\text{int}} + b_{\text{int}\,i} + (\beta_{\text{cov}} + b_{\text{cov}\,i})x_{i,t}\}. \qquad (2.2.9)$$

The asymptote depends on the covariate $x_{i,t}$. The term $(1 - \rho)^{-1}b_{\text{cov}\,i}$ represents the difference in sensitivity to dose modifications across subjects. In some subjects, the response level will change largely according to changes in dosing; in other subjects, the response level will not change according to changes in dosing. In Sects. 3.3 and 4.3, examples of clinical studies in which the treatment dose is a time-dependent covariate are shown.

2.3 Autoregressive Linear Mixed Effects Models

In this section, we introduce autoregressive linear mixed effects models (Funatogawa et al. 2007, 2008a; Funatogawa et al. 2008b; Funatogawa and Funatogawa 2012a, b). Similar to the example in Sect. 2.2.1, the model is represented in three ways: an autoregressive form, response changes with asymptotes, and a marginal (unconditional) form. These representations are summarized in Table 2.1, along with the nonlinear mixed effects models and differential equations in Chap. 5 and the state space form in Chap. 6.

2.3.1 Autoregressive Form

Let $\mathbf{Y}_i = \left(Y_{i,0}, Y_{i,1}, Y_{i,2}, \cdots, Y_{i,T_i}\right)^T$ be the $(T_i + 1) \times 1$ vector of the response corresponding to the ith $(i = 1, \cdots, N)$ subject measured from 0 to T_i. $Y_{i,0}$ is a baseline measurement, and $Y_{i,t}(t = 1, 2, \cdots, T_i)$ is the tth measurement after the baseline measurement. $\mathbf{A}^T$ denotes the transpose of a matrix $\mathbf{A}$. For the vector representation, we introduce a $(T_i + 1) \times (T_i + 1)$ matrix $\mathbf{F}_i$ whose elements just below the diagonal are 1 and the other elements are 0. Then, $\mathbf{F}_i \mathbf{Y}_i$ is the vector of the previous response as

$$\mathbf{F}_i \mathbf{Y}_i = \left(0, Y_{i,0}, Y_{i,1}, \cdots, Y_{i,T_i-1}\right)^T. \qquad (2.3.1)$$

For $T_i = 3$, $\mathbf{F}_i$ and $\mathbf{F}_i \mathbf{Y}_i$ are

Table 2.1 Representations of autoregressive linear mixed effects models

Representation	
(a) Autoregressive form	$\mathbf{Y}_i = \rho \mathbf{F}_i \mathbf{Y}_i + \mathbf{X}_i \boldsymbol{\beta} + \mathbf{Z}_i \mathbf{b}_i + \boldsymbol{\varepsilon}_i$
	$\mathbf{V}_i = \text{Var}(\mathbf{Z}_i \mathbf{b}_i + \boldsymbol{\varepsilon}_i) = \mathbf{Z}_i \mathbf{G} \mathbf{Z}_i^T + \mathbf{R}_i$
(b) Response changes with asymptotes[a]	$\begin{cases} \mathbf{Y}_i - \mathbf{F}_i \mathbf{Y}_i = \mathbf{J}_i \left(\mathbf{Y}_{\text{Base Asy } i} - \mathbf{F}_i \mathbf{Y}_i \right) + \boldsymbol{\varepsilon}_i \\ \mathbf{Y}_{\text{Base Asy } i} = \mathbf{X}_i \boldsymbol{\beta}^* + \mathbf{Z}_i \mathbf{b}_i^* \\ \mathbf{b}_i^* \sim \text{MVN}\left(\mathbf{0}, \mathbf{M}_z \mathbf{G} \mathbf{M}_z^T \right) \\ \boldsymbol{\varepsilon}_i \sim \text{MVN}(\mathbf{0}, \mathbf{R}_i) \end{cases}$ where $\boldsymbol{\beta}^* = \mathbf{M}_x \boldsymbol{\beta}$ and $\mathbf{b}_i^* = \mathbf{M}_z \mathbf{b}_i$
(c) Marginal (unconditional) form	$\mathbf{Y}_i = (\mathbf{I}_i - \rho \mathbf{F}_i)^{-1} (\mathbf{X}_i \boldsymbol{\beta} + \mathbf{Z}_i \mathbf{b}_i + \boldsymbol{\varepsilon}_i)$ $\boldsymbol{\Sigma}_i = \text{Var}(\mathbf{Y}_i)$ $\quad = \text{Var}\left\{ (\mathbf{I}_i - \rho \mathbf{F}_i)^{-1} (\mathbf{Z}_i \mathbf{b}_i + \boldsymbol{\varepsilon}_i) \right\}$ $\quad = (\mathbf{I}_i - \rho \mathbf{F}_i)^{-1} \left(\mathbf{Z}_i \mathbf{G} \mathbf{Z}_i^T + \mathbf{R}_i \right) \left\{ (\mathbf{I}_i - \rho \mathbf{F}_i)^{-1} \right\}^T$ $\quad = (\mathbf{I}_i - \rho \mathbf{F}_i)^{-1} \mathbf{V}_i \left\{ (\mathbf{I}_i - \rho \mathbf{F}_i)^{-1} \right\}^T$
(d) Nonlinear mixed effects models (without covariate)[b]	$\begin{cases} Y_{ij} = f\left(t_{ij}, \boldsymbol{\beta}, \mathbf{b}_i \right) + \varepsilon_{ij} \\ f\left(t_{ij}, \boldsymbol{\beta}, \mathbf{b}_i \right) = (\beta_1 + b_{1i}) e^{-\beta_3 t_{ij}} + (\beta_2 + b_{2i})\left(1 - e^{-\beta_3 t_{ij}} \right) \\ \mathbf{b}_i = (b_{1i}, b_{2i})^T, \mathbf{b}_i \sim \text{MVN}(\mathbf{0}, \mathbf{G}) \\ \boldsymbol{\varepsilon}_i = \left(\varepsilon_{i1}, \cdots, \varepsilon_{in_i} \right)^T, \boldsymbol{\varepsilon}_i \sim \text{MVN}(\mathbf{0}, \mathbf{R}_i) \end{cases}$
(e) Differential equation (with a time-dependent covariate)[b]	$\begin{cases} d\mu_i(t)/dt = \kappa \{ \beta_2 + b_{2i} + (\beta_c + b_{ci}) x_i(t) - \mu_i(t) \} \\ \mu_i(0) = \beta_1 + b_{1i} \\ Y_{ij} = \mu_i(t_{ij}) + \varepsilon_{ij} \\ \mathbf{b}_i = (b_{1i}, b_{2i}, b_{ci})^T, \mathbf{b}_i \sim \text{MVN}(\mathbf{0}, \mathbf{G}) \\ \boldsymbol{\varepsilon}_i = \left(\varepsilon_{i1}, \cdots, \varepsilon_{in_i} \right)^T, \boldsymbol{\varepsilon}_i \sim \text{MVN}(\mathbf{0}, \mathbf{R}_i) \end{cases}$
(f) State space[c]	$\begin{pmatrix} \mu_{i,t} \\ \mathbf{b}_i \end{pmatrix} = \begin{pmatrix} \rho & \mathbf{Z}_{i,t} \\ \mathbf{0}_{q \times 1} & \mathbf{I}_{q \times q} \end{pmatrix} \begin{pmatrix} \mu_{i,t-1} \\ \mathbf{b}_i \end{pmatrix} + \begin{pmatrix} \mathbf{X}_{i,t} \boldsymbol{\beta} \\ \mathbf{0}_{q \times 1} \end{pmatrix} + \begin{pmatrix} \varepsilon_{(\text{AR})i,t} \\ \mathbf{0}_{q \times 1} \end{pmatrix}$ $Y_{i,t} = \begin{pmatrix} 1 & \mathbf{0}_{1 \times q} \end{pmatrix} \begin{pmatrix} \mu_{i,t} \\ \mathbf{b}_i \end{pmatrix} + \varepsilon_{(\text{ME})i,t}$ $\mathbf{s}_{i(-1|-1)} = \mathbf{0}_{(1+q) \times 1}$ $\mathbf{Q}_{i,0} \equiv \text{Var} \begin{pmatrix} \varepsilon_{(\text{AR})i,0} \\ \mathbf{0}_{q \times 1} \end{pmatrix} = \mathbf{0}_{(q+1) \times (q+1)}, \mathbf{Q}_{i,t} = \begin{pmatrix} \sigma_{\text{AR}}^2 & \mathbf{0}_{1 \times q} \\ \mathbf{0}_{q \times 1} & \mathbf{0}_{q \times q} \end{pmatrix}$ $r_{i,t} \equiv \text{Var}\left(\varepsilon_{(\text{ME})i,t} \right) = \sigma_{\text{ME}}^2, \mathbf{P}_{i(-1|-1)} = \begin{pmatrix} 0_{1 \times 1} & \mathbf{0}_{1 \times q} \\ \mathbf{0}_{q \times 1} & \mathbf{G} \end{pmatrix}$

[a] See Sect. 2.3.2 for the definition of $\mathbf{J}_i$, $\mathbf{M}_x$, and $\mathbf{M}_z$. [b] See Chap. 5 for more details of the nonlinear mixed effects models and differential equations. [c] See Chap. 6 for more details of the state space representation

$$\mathbf{F}_i = \begin{pmatrix} 0 & 0 & 0 & 0 \\ 1 & 0 & 0 & 0 \\ 0 & 1 & 0 & 0 \\ 0 & 0 & 1 & 0 \end{pmatrix},$$ (2.3.2)

$$\mathbf{F}_i\mathbf{Y}_i = \begin{pmatrix} 0 & 0 & 0 & 0 \\ 1 & 0 & 0 & 0 \\ 0 & 1 & 0 & 0 \\ 0 & 0 & 1 & 0 \end{pmatrix} \begin{pmatrix} Y_{i,0} \\ Y_{i,1} \\ Y_{i,2} \\ Y_{i,3} \end{pmatrix} = \begin{pmatrix} 0 \\ Y_{i,0} \\ Y_{i,1} \\ Y_{i,2} \end{pmatrix}.$$

Autoregressive linear mixed effects models are expressed as

$$\mathbf{Y}_i = \rho\mathbf{F}_i\mathbf{Y}_i + \mathbf{X}_i\boldsymbol{\beta} + \mathbf{Z}_i\mathbf{b}_i + \boldsymbol{\varepsilon}_i,$$ (2.3.3)

where ρ is an unknown regression coefficient of the previous response, $\boldsymbol{\beta}$ is a $p \times 1$ vector of unknown fixed effects parameters, $\mathbf{X}_i$ is a known $(T_i + 1) \times p$ design matrix of fixed effects, $\mathbf{b}_i$ is a $q \times 1$ vector of unknown random effects parameters, $\mathbf{Z}_i$ is a known $(T_i + 1) \times q$ design matrix of random effects, and $\boldsymbol{\varepsilon}_i$ is a $(T_i + 1) \times 1$ vector of random errors. It is assumed that $\mathbf{b}_i$ and $\boldsymbol{\varepsilon}_i$ are both independent across subjects and independently normally distributed with the mean zero vector and the variance covariance matrices $\mathbf{G}$ and $\mathbf{R}_i$, respectively,

$$\mathbf{b}_i \sim \text{MVN}(\mathbf{0}, \mathbf{G}),$$ (2.3.4)

$$\boldsymbol{\varepsilon}_i \sim \text{MVN}(\mathbf{0}, \mathbf{R}_i).$$ (2.3.5)

Let $\mathbf{V}_i$ be the variance covariance matrix of the response vector $\mathbf{Y}_i$ conditional on the previous response $\mathbf{F}_i\mathbf{Y}_i$. As with the linear mixed effects models shown in Sect. 1.2, the variance covariance matrix is written as

$$\mathbf{V}_i = \text{Var}(\mathbf{Z}_i\mathbf{b}_i + \boldsymbol{\varepsilon}_i)$$
$$= \mathbf{Z}_i\mathbf{G}\mathbf{Z}_i^T + \mathbf{R}_i.$$ (2.3.6)

The explicit difference from linear mixed effects models (1.2.1) is the inclusion of the term $\rho\mathbf{F}_i\mathbf{Y}_i$. However, there are also differences in the interpretation of model parameters, changes in the response level, and variance covariance structures. We will see these points in the following sections.

Table 2.2a gives the vector representation of the model (2.2.8) of autoregressive linear mixed effects models with a time-dependent covariate for $T_i = 3$. In this example, $\mathbf{X}_i = \mathbf{Z}_i$. These are block diagonal matrices: the blocks correspond to the baseline parts ($t = 0$) and the other parts ($t > 0$).

Table 2.2 Three representations of an example of autoregressive linear mixed effects models for $T_i = 3$

Representation

(a) Autoregressive form $\mathbf{Y}_i = \rho \mathbf{F}_i \mathbf{Y}_i + \mathbf{X}_i \boldsymbol{\beta} + \mathbf{Z}_i \mathbf{b}_i + \boldsymbol{\varepsilon}_i$ (2.3.3). Model (2.2.8)

$$
\begin{pmatrix} Y_{i,0} \\ Y_{i,1} \\ Y_{i,2} \\ Y_{i,3} \end{pmatrix} = \rho \begin{pmatrix} 0 \\ Y_{i,0} \\ Y_{i,1} \\ Y_{i,2} \end{pmatrix} + \begin{pmatrix} 1 & 0 & 0 \\ 0 & 1 & x_{i,1} \\ 0 & 1 & x_{i,2} \\ 0 & 1 & x_{i,3} \end{pmatrix} \begin{pmatrix} \beta_{\text{base}} \\ \beta_{\text{int}} \\ \beta_{\text{cov}} \end{pmatrix} + \begin{pmatrix} 1 & 0 & 0 \\ 0 & 1 & x_{i,1} \\ 0 & 1 & x_{i,2} \\ 0 & 1 & x_{i,3} \end{pmatrix} \begin{pmatrix} b_{\text{base}\,i} \\ b_{\text{int}\,i} \\ b_{\text{cov}\,i} \end{pmatrix} + \begin{pmatrix} \varepsilon_{i,0} \\ \varepsilon_{i,1} \\ \varepsilon_{i,2} \\ \varepsilon_{i,3} \end{pmatrix}
$$

$$
= \begin{pmatrix} 0 \\ \rho Y_{i,0} \\ \rho Y_{i,1} \\ \rho Y_{i,2} \end{pmatrix} + \begin{pmatrix} \beta_{\text{base}} + b_{\text{base}\,i} \\ \beta_{\text{int}} + b_{\text{int}\,i} + (\beta_{\text{cov}} + b_{\text{cov}\,i}) x_{i,1} \\ \beta_{\text{int}} + b_{\text{int}\,i} + (\beta_{\text{cov}} + b_{\text{cov}\,i}) x_{i,2} \\ \beta_{\text{int}} + b_{\text{int}\,i} + (\beta_{\text{cov}} + b_{\text{cov}\,i}) x_{i,3} \end{pmatrix} + \begin{pmatrix} \varepsilon_{i,0} \\ \varepsilon_{i,1} \\ \varepsilon_{i,2} \\ \varepsilon_{i,3} \end{pmatrix}
$$

where $\mathbf{F}_i = \begin{pmatrix} 0 & 0 & 0 & 0 \\ 1 & 0 & 0 & 0 \\ 0 & 1 & 0 & 0 \\ 0 & 0 & 1 & 0 \end{pmatrix}$ and $\mathbf{F}_i \mathbf{Y}_i = \begin{pmatrix} 0 \\ Y_{i,0} \\ Y_{i,1} \\ Y_{i,2} \end{pmatrix}$

(b) Response changes with asymptotes $\mathbf{Y}_i - \mathbf{F}_i \mathbf{Y}_i = \mathbf{J}_i \left(\mathbf{Y}_{\text{Base Asy}\,i} - \mathbf{F}_i \mathbf{Y}_i \right) + \boldsymbol{\varepsilon}_i$ with
$\mathbf{Y}_{\text{Base Asy}\,i} = \mathbf{X}_i \mathbf{M}_x \boldsymbol{\beta} + \mathbf{Z}_i \mathbf{M}_z \mathbf{b}_i = \mathbf{X}_i \boldsymbol{\beta}^* + \mathbf{Z}_i \mathbf{b}_i^*$ (2.3.13) where *(asterisk) shows the parameters for the asymptote. Model (2.2.8) with the representation (2.3.7)

$$
\begin{pmatrix} Y_{i,0} \\ Y_{i,1} - Y_{i,0} \\ Y_{i,2} - Y_{i,1} \\ Y_{i,3} - Y_{i,2} \end{pmatrix} = \begin{pmatrix} 1 & 0 & 0 & 0 \\ 0 & 1-\rho & 0 & 0 \\ 0 & 0 & 1-\rho & 0 \\ 0 & 0 & 0 & 1-\rho \end{pmatrix} \left[\begin{pmatrix} 1 & 0 & 0 \\ 0 & 1 & x_{i,1} \\ 0 & 1 & x_{i,2} \\ 0 & 1 & x_{i,3} \end{pmatrix} \begin{pmatrix} 1 & 0 & 0 \\ 0 & (1-\rho)^{-1} & 0 \\ 0 & 0 & (1-\rho)^{-1} \end{pmatrix} \begin{pmatrix} \beta_{\text{base}} \\ \beta_{\text{int}} \\ \beta_{\text{cov}} \end{pmatrix} \right.
$$

$$
\left. + \begin{pmatrix} 1 & 0 & 0 \\ 0 & 1 & x_{i,1} \\ 0 & 1 & x_{i,2} \\ 0 & 1 & x_{i,3} \end{pmatrix} \begin{pmatrix} 1 & 0 & 0 \\ 0 & (1-\rho)^{-1} & 0 \\ 0 & 0 & (1-\rho)^{-1} \end{pmatrix} \begin{pmatrix} b_{\text{base}\,i} \\ b_{\text{int}\,i} \\ b_{\text{cov}\,i} \end{pmatrix} - \begin{pmatrix} 0 \\ Y_{i,0} \\ Y_{i,1} \\ Y_{i,2} \end{pmatrix} \right] + \begin{pmatrix} \varepsilon_{i,0} \\ \varepsilon_{i,1} \\ \varepsilon_{i,2} \\ \varepsilon_{i,3} \end{pmatrix}
$$

$$
= \begin{pmatrix} 1 & 0 & 0 & 0 \\ 0 & 1-\rho & 0 & 0 \\ 0 & 0 & 1-\rho & 0 \\ 0 & 0 & 0 & 1-\rho \end{pmatrix} \left[\begin{pmatrix} 1 & 0 & 0 \\ 0 & 1 & x_{i,1} \\ 0 & 1 & x_{i,2} \\ 0 & 1 & x_{i,3} \end{pmatrix} \begin{pmatrix} \beta_{\text{base}} + b_{\text{base}\,i} \\ \beta_{\text{int}}^* + b_{\text{int}\,i}^* \\ \beta_{\text{cov}}^* + b_{\text{cov}\,i}^* \end{pmatrix} - \begin{pmatrix} 0 \\ Y_{i,0} \\ Y_{i,1} \\ Y_{i,2} \end{pmatrix} \right] + \begin{pmatrix} \varepsilon_{i,0} \\ \varepsilon_{i,1} \\ \varepsilon_{i,2} \\ \varepsilon_{i,3} \end{pmatrix}
$$

$$
= \begin{pmatrix} \beta_{\text{base}} + b_{\text{base}\,i} \\ (1-\rho)\{\beta_{\text{int}}^* + b_{\text{int}\,i}^* + (\beta_{\text{cov}}^* + b_{\text{cov}\,i}^*) x_{i,1} - Y_{i,0}\} \\ (1-\rho)\{\beta_{\text{int}}^* + b_{\text{int}\,i}^* + (\beta_{\text{cov}}^* + b_{\text{cov}\,i}^*) x_{i,2} - Y_{i,1}\} \\ (1-\rho)\{\beta_{\text{int}}^* + b_{\text{int}\,i}^* + (\beta_{\text{cov}}^* + b_{\text{cov}\,i}^*) x_{i,3} - Y_{i,2}\} \end{pmatrix} + \begin{pmatrix} \varepsilon_{i,0} \\ \varepsilon_{i,1} \\ \varepsilon_{i,2} \\ \varepsilon_{i,3} \end{pmatrix}
$$

where $\mathbf{J}_i = \begin{pmatrix} 1 & 0 & 0 & 0 \\ 0 & 1-\rho & 0 & 0 \\ 0 & 0 & 1-\rho & 0 \\ 0 & 0 & 0 & 1-\rho \end{pmatrix}$ and $\mathbf{M}_x = \mathbf{M}_z = \begin{pmatrix} 1 & 0 & 0 \\ 0 & (1-\rho)^{-1} & 0 \\ 0 & 0 & (1-\rho)^{-1} \end{pmatrix}$

(continued)

Table 2.2 (continued)

Representation

(c) Marginal (unconditional) form $\mathbf{Y}_i = (\mathbf{I}_i - \rho\mathbf{F}_i)^{-1}(\mathbf{X}_i\boldsymbol{\beta} + \mathbf{Z}_i\mathbf{b}_i + \boldsymbol{\varepsilon}_i)$ (2.3.14).
Model (2.2.8) with the marginal form (2.3.19)

$$
\begin{pmatrix} Y_{i,0} \\ Y_{i,1} \\ Y_{i,2} \\ Y_{i,3} \end{pmatrix} = \begin{pmatrix} 1 & 0 & 0 & 0 \\ \rho & 1 & 0 & 0 \\ \rho^2 & \rho & 1 & 0 \\ \rho^3 & \rho^2 & \rho & 1 \end{pmatrix} \left\{ \begin{pmatrix} 1 & 0 & 0 \\ 0 & 1 & x_{i,1} \\ 0 & 1 & x_{i,2} \\ 0 & 1 & x_{i,3} \end{pmatrix} \begin{pmatrix} \beta_{\text{base}} \\ \beta_{\text{int}} \\ \beta_{\text{cov}} \end{pmatrix} + \begin{pmatrix} 1 & 0 & 0 \\ 0 & 1 & x_{i,1} \\ 0 & 1 & x_{i,2} \\ 0 & 1 & x_{i,3} \end{pmatrix} \begin{pmatrix} b_{\text{base }i} \\ b_{\text{int }i} \\ b_{\text{cov }i} \end{pmatrix} + \begin{pmatrix} \varepsilon_{i,0} \\ \varepsilon_{i,1} \\ \varepsilon_{i,2} \\ \varepsilon_{i,3} \end{pmatrix} \right\}
$$

where $(\mathbf{I}_i - \rho\mathbf{F}_i)^{-1} = \begin{pmatrix} 1 & 0 & 0 & 0 \\ \rho & 1 & 0 & 0 \\ \rho^2 & \rho & 1 & 0 \\ \rho^3 & \rho^2 & \rho & 1 \end{pmatrix}$

2.3.2 Representation of Response Changes with Asymptotes

The model (2.2.8) can be represented by response changes with asymptotes as

$$
\begin{cases}
Y_{i,0} = \beta_{\text{base}} + b_{\text{base }i} + \varepsilon_{i,0} \\
Y_{i,t} - Y_{i,t-1} = (1-\rho)\left(Y_{\text{Asy }i,t} - Y_{i,t-1}\right) + \varepsilon_{i,t}, \quad (t > 0) \\
Y_{\text{Asy }i,t} = (1-\rho)^{-1}\left\{ \beta_{\text{int}} + b_{\text{int }i} + (\beta_{\text{cov}} + b_{\text{cov }i})x_{i,t} \right\} \\
\qquad\quad = \beta_{\text{int}}^* + b_{\text{int }i}^* - \left(\beta_{\text{cov}}^* + b_{\text{cov }i}^*\right)x_{i,t}
\end{cases} \tag{2.3.7}
$$

where * (asterisk) shows the parameters for the asymptote. The asymptote linearly depends on the covariate.

Changes at each time point can be shown in the following vector representation:

$$
\begin{cases}
Y_{i,0} = \mathbf{X}_{i,0}\boldsymbol{\beta} + \mathbf{Z}_{i,0}\mathbf{b}_i + \boldsymbol{\varepsilon}_{i,0} \\
Y_{i,t} - Y_{i,t-1} = (1-\rho)\left(Y_{\text{Asy }i,t} - Y_{i,t-1}\right) + \varepsilon_{i,t}, \quad (t > 0) \\
Y_{\text{Asy }i,t} = (1-\rho)^{-1}\left(\mathbf{X}_{i,t}\boldsymbol{\beta} + \mathbf{Z}_{i,t}\mathbf{b}_i\right) \\
\qquad\quad = \mathbf{X}_{i,t}\boldsymbol{\beta}^* + \mathbf{Z}_{i,t}\mathbf{b}_i^*
\end{cases} \tag{2.3.8}
$$

where $\mathbf{X}_{i,t}$ and $\mathbf{Z}_{i,t}$ are the corresponding $1 \times p$ and $1 \times q$ row vectors of $\mathbf{X}_i$ and $\mathbf{Z}_i$. We consider only the case that $\mathbf{X}_i$ and $\mathbf{Z}_i$ are both block diagonal matrices and the blocks correspond to the baseline parts ($t = 0$) and the other parts ($t > 0$). The parameters βs and bs are transformed into the new parameters β^*s and b^*s for the asymptote by multiplying $(1 - \rho)^{-1}$. For the vector representations of the parameter transformations $\boldsymbol{\beta}^* = \mathbf{M}_x\boldsymbol{\beta}$ and $\mathbf{b}_i^* = \mathbf{M}_z\mathbf{b}_i$, we introduce a $p \times p$ diagonal matrix $\mathbf{M}_x$ and a $q \times q$ diagonal matrix $\mathbf{M}_z$. The diagonal elements of $\mathbf{M}_x$ and $\mathbf{M}_z$ are 1 for the baseline parameters, and $(1 - \rho)^{-1}$ for the parameters of the later time points. For example, there are three fixed effects parameters, β_{base}, β_{int}, β_{cov}, in the model (2.3.7), and $\mathbf{M}_x$ is

$$\mathbf{M}_x = \begin{pmatrix} 1 & 0 & 0 \\ 0 & (1-\rho)^{-1} & 0 \\ 0 & 0 & (1-\rho)^{-1} \end{pmatrix}. \tag{2.3.9}$$

Changes at all the time points can be shown in the following vector representation:

$$\mathbf{Y}_i - \mathbf{F}_i \mathbf{Y}_i = \mathbf{J}_i (\mathbf{X}_i \mathbf{M}_x \boldsymbol{\beta} + \mathbf{Z}_i \mathbf{M}_z \mathbf{b}_i - \mathbf{F}_i \mathbf{Y}_i) + \boldsymbol{\varepsilon}_i. \tag{2.3.10}$$

$\mathbf{Y}_i - \mathbf{F}_i \mathbf{Y}_i$ is a $(T_i + 1) \times 1$ vector in which the first element is a baseline response, $Y_{i,0}$, and the other elements are response changes, $Y_{i,t} - Y_{i,t-1}$. For the vector representation, we introduce a $(T_i + 1) \times (T_i + 1)$ diagonal matrix $\mathbf{J}_i$. The $(1, 1)$th element of $\mathbf{J}_i$ is 1, which corresponds to the baseline $(t = 0)$, and the other elements are the proportional constant $(1 - \rho)$ which corresponds to the later time points $(t > 0)$. Thus, for $T_i = 3$, $\mathbf{J}_i$ is

$$\mathbf{J}_i = \begin{pmatrix} 1 & 0 & 0 & 0 \\ 0 & 1-\rho & 0 & 0 \\ 0 & 0 & 1-\rho & 0 \\ 0 & 0 & 0 & 1-\rho \end{pmatrix}. \tag{2.3.11}$$

Let $\mathbf{Y}_{\text{Base Asy } i}$ be the $(T_i + 1) \times 1$ vector, and the first element corresponds to the baseline, and the other elements correspond to the asymptotes,

$$\begin{aligned} \mathbf{Y}_{\text{Base Asy } i} &\equiv \mathbf{X}_i \mathbf{M}_x \boldsymbol{\beta} + \mathbf{Z}_i \mathbf{M}_z \mathbf{b}_i \\ &= \mathbf{X}_i \boldsymbol{\beta}^* + \mathbf{Z}_i \mathbf{b}_i^*. \end{aligned} \tag{2.3.12}$$

Then, the representation of response changes with asymptotes of autoregressive linear mixed effects models (2.3.3) is

$$\begin{cases} \mathbf{Y}_i - \mathbf{F}_i \mathbf{Y}_i = \mathbf{J}_i \left(\mathbf{Y}_{\text{Base Asy } i} - \mathbf{F}_i \mathbf{Y}_i \right) + \boldsymbol{\varepsilon}_i \\ \mathbf{Y}_{\text{Base Asy } i} = \mathbf{X}_i \boldsymbol{\beta}^* + \mathbf{Z}_i \mathbf{b}_i^* \end{cases}. \tag{2.3.13}$$

The expected value of $\mathbf{Y}_{\text{Base Asy } i}$ is

$$\begin{aligned} \mathrm{E}\left(\mathbf{Y}_{\text{Base Asy } i} \right) &= \mathrm{E}(\mathbf{X}_i \mathbf{M}_x \boldsymbol{\beta} + \mathbf{Z}_i \mathbf{M}_z \mathbf{b}_i) \\ &= \mathbf{X}_i \mathbf{M}_x \boldsymbol{\beta} \\ &= \mathbf{X}_i \boldsymbol{\beta}^*. \end{aligned}$$

The expected change from $Y_{i,t-1}$ to $Y_{i,t}$ given random effects is proportional to the size remaining to the asymptote, $\mathbf{Y}_{\text{Base Asy } i} - \mathbf{F}_i \mathbf{Y}_i$ except for the first element. Table 2.2b gives the vector representation of (2.3.7) for $T_i = 3$.

2.3.3 Marginal Form

The marginal form (unconditional form) of autoregressive linear mixed effects models (2.3.3) is

$$\mathbf{Y}_i = (\mathbf{I}_i - \rho\mathbf{F}_i)^{-1}(\mathbf{X}_i\boldsymbol{\beta} + \mathbf{Z}_i\mathbf{b}_i + \boldsymbol{\varepsilon}_i), \tag{2.3.14}$$

where $\mathbf{I}_i$ is a $(T_i + 1) \times (T_i + 1)$ identity matrix. In this section, the subscript i of $\mathbf{I}_i$ indicates the subject i instead of the size of the identity matrix. The above equation is derived from multiplying both sides of the following equation by $(\mathbf{I}_i - \rho\mathbf{F}_i)^{-1}$,

$$\mathbf{Y}_i - \rho\mathbf{F}_i\mathbf{Y}_i = \mathbf{X}_i\boldsymbol{\beta} + \mathbf{Z}_i\mathbf{b}_i + \boldsymbol{\varepsilon}_i. \tag{2.3.15}$$

For $T_i = 3$, $(\mathbf{I}_i - \rho\mathbf{F}_i)^{-1}$ is

$$(\mathbf{I}_i - \rho\mathbf{F}_i)^{-1} = \begin{pmatrix} 1 & 0 & 0 & 0 \\ -\rho & 1 & 0 & 0 \\ 0 & -\rho & 1 & 0 \\ 0 & 0 & -\rho & 1 \end{pmatrix}^{-1} = \begin{pmatrix} 1 & 0 & 0 & 0 \\ \rho & 1 & 0 & 0 \\ \rho^2 & \rho & 1 & 0 \\ \rho^3 & \rho^2 & \rho & 1 \end{pmatrix}. \tag{2.3.16}$$

The expectation given random effects $\mathbf{b}_i$ is $E(\mathbf{Y}_i|\mathbf{b}_i) = (\mathbf{I}_i - \rho\mathbf{F}_i)^{-1}(\mathbf{X}_i\boldsymbol{\beta} + \mathbf{Z}_i\mathbf{b}_i)$, and the marginal expectation is $E(\mathbf{Y}_i) = (\mathbf{I}_i - \rho\mathbf{F}_i)^{-1}\mathbf{X}_i\boldsymbol{\beta}$. The expectation for a typical subject and the marginal expectation are the same, $E(\mathbf{Y}_i|\mathbf{b}_i = \mathbf{0}) = E(\mathbf{Y}_i)$. Subject specific interpretation and marginal interpretation are the same. Let $\boldsymbol{\Sigma}_i$ be the marginal variance covariance matrix of the response vector $\mathbf{Y}_i$. $\boldsymbol{\Sigma}_i$ is

$$\begin{aligned} \boldsymbol{\Sigma}_i &= \mathrm{Var}(\mathbf{Y}_i) \\ &= \mathrm{Var}\{(\mathbf{I}_i - \rho\mathbf{F}_i)^{-1}(\mathbf{Z}_i\mathbf{b}_i + \boldsymbol{\varepsilon}_i)\} \\ &= (\mathbf{I}_i - \rho\mathbf{F}_i)^{-1}(\mathbf{Z}_i\mathbf{G}\mathbf{Z}_i^T + \mathbf{R}_i)\{(\mathbf{I}_i - \rho\mathbf{F}_i)^{-1}\}^T \\ &= (\mathbf{I}_i - \rho\mathbf{F}_i)^{-1}\mathbf{V}_i\{(\mathbf{I}_i - \rho\mathbf{F}_i)^{-1}\}^T. \end{aligned} \tag{2.3.17}$$

Given the marginal variance covariance matrix $\boldsymbol{\Sigma}_i$, the variance covariance matrix of the autoregressive form $\mathbf{V}_i$ can be expressed as

$$\mathbf{V}_i = (\mathbf{I}_i - \rho\mathbf{F}_i)\boldsymbol{\Sigma}_i(\mathbf{I}_i - \rho\mathbf{F}_i)^T. \tag{2.3.18}$$

The marginal form of the model (2.2.8) is

$$\begin{cases} Y_{i,0} = \beta_{\text{base}} + b_{\text{base }i} + \varepsilon_{\text{m }i,0} \\ Y_{i,t} = \rho^t(\beta_{\text{base}} + b_{\text{base }i}) \\ \qquad + \sum_{l=1}^{t} \rho^{t-l}\{\beta_{\text{int}} + b_{\text{int }i} + (\beta_{\text{cov}} + b_{\text{cov }i})x_{i,l}\} + \varepsilon_{\text{m }i,t}, \ (t > 0) \end{cases}, \qquad (2.3.19)$$

where $\varepsilon_{\text{m }i,t} = \sum_{l=0}^{t} \rho^{t-l}\varepsilon_{i,l}$. Table 2.2c provides the vector representation of (2.3.19) for $T_i = 3$.

2.4 Variance Covariance Structures

This section shows the variance covariance structures in detail. The variance covariance matrix of the response vector in autoregressive linear mixed effects models is $\mathbf{V}_i = \mathbf{Z}_i\mathbf{G}\mathbf{Z}_i^T + \mathbf{R}_i$ as (2.3.6). The marginal form is $\mathbf{\Sigma}_i = (\mathbf{I}_i - \rho\mathbf{F}_i)^{-1}(\mathbf{Z}_i\mathbf{G}\mathbf{Z}_i^T + \mathbf{R}_i)\{(\mathbf{I}_i - \rho\mathbf{F}_i)^{-1}\}^T$ as (2.3.17). This matrix has two parts, $(\mathbf{I}_i - \rho\mathbf{F}_i)^{-1}(\mathbf{Z}_i\mathbf{G}\mathbf{Z}_i^T)\{(\mathbf{I}_i - \rho\mathbf{F}_i)^{-1}\}^T$ and $(\mathbf{I}_i - \rho\mathbf{F}_i)^{-1}\mathbf{R}_i\{(\mathbf{I}_i - \rho\mathbf{F}_i)^{-1}\}^T$. Given $\mathbf{\Sigma}_i$, $\mathbf{V}_i = (\mathbf{I}_i - \rho\mathbf{F}_i)\mathbf{\Sigma}_i(\mathbf{I}_i - \rho\mathbf{F}_i)^T$ as (2.3.18). Section 2.4.1 shows structures of random errors, in particular, the AR(1) error and measurement error, and the variance covariance matrices $\mathbf{R}_i$ which are induced by the AR(1) error and measurement error. Section 2.4.2 shows the structure of $\mathbf{Z}_i\mathbf{G}\mathbf{Z}_i^T$ which is induced by random effects. Section 2.4.3 shows the variance covariance matrices $\mathbf{V}_i$ and $\mathbf{\Sigma}_i$ which are induced by both random effects and random errors. Section 2.4.4 shows the variance covariance matrix for asymptotes.

2.4.1 AR(1) Error and Measurement Error

In linear mixed effects models with random effects, an independent error is often assumed for random errors. However, an independent error in an autoregressive form is an AR(1) error in the marginal form. In actual data analysis, a measurement error that is independent across time is often seen. An independent error in a marginal form is a reasonable assumption, particularly when the measurement method is imprecise. We consider the error structures induced by an AR(1) error and a measurement error simultaneously. We consider two different assumptions for an AR(1) error term in $Y_{i,0}$.

First, $\varepsilon_{i,0}$ does not include an AR(1) error. The error structure in the autoregressive form such as (2.2.1) is

$$\begin{cases} \varepsilon_{i,0} = \varepsilon_{(\text{ME})i,0} \\ \varepsilon_{i,t} = \varepsilon_{(\text{AR})i,t} + \varepsilon_{(\text{ME})i,t} - \rho\varepsilon_{(\text{ME})i,t-1}, \ (t > 0) \end{cases}, \qquad (2.4.1)$$

where $\varepsilon_{(AR)i,t}$ and $\varepsilon_{(ME)i,t}$ independently follow a normal distribution with the mean 0 and the variances σ^2_{AR} and σ^2_{ME}, respectively. Here, AR means autoregressive, and ME means a measurement error. In the marginal form such as (2.2.3), this error structure is

$$\begin{cases} \varepsilon_{m\,i,0} = \varepsilon_{(ME)i,0} \\ \varepsilon_{m\,i,t} = \sum_{j=1}^{t} \rho^{t-j} \varepsilon_{(AR)i,j} + \varepsilon_{(ME)i,t}, \, (t > 0) \end{cases}$$
(2.4.2)

From these equations, we can confirm that $\varepsilon_{(ME)i,t}$ is an independent error in the marginal form. The variance of $\varepsilon_{m\,i,t}$ is not constant and this AR(1) error is not stationary.

Second, we consider a stationary AR(1) error. In the autoregressive form, the error structure is

$$\begin{cases} \varepsilon_{i,0} = \varepsilon_{(AR,ST)i,0} + \varepsilon_{(ME)i,0} \\ \varepsilon_{i,t} = \varepsilon_{(AR,ST)i,t} + \varepsilon_{(ME)i,t} - \rho \varepsilon_{(ME)i,t-1}, \, (t > 0) \end{cases}$$
(2.4.3)

where $\varepsilon_{(AR,ST)i,0}$, $\varepsilon_{(AR,ST)i,t}$ $(t > 0)$, and $\varepsilon_{(ME)i,t}$ independently follow a normal distribution with the mean 0 and the variances $(1 - \rho^2)^{-1} \sigma^2_{AR,ST}$, $\sigma^2_{AR,ST}$, and σ^2_{ME}, respectively. Here, ST in subscript means stationary. In the marginal form, this error structure is

$$\varepsilon_{m\,i,t} = \sum_{j=0}^{t} \rho^{t-j} \varepsilon_{(AR,ST)i,j} + \varepsilon_{(ME)i,t}.$$
(2.4.4)

The variance of $\varepsilon_{m\,i,t}$ is $\text{Var}(\varepsilon_{m\,i,t}) = (1 - \rho^2)^{-1} \sigma^2_{AR,ST} + \sigma^2_{ME}$ and it is constant.
The model (2.2.1) $(t > 0)$ with the error structure (2.4.1) is

$$Y_{i,t} = \rho Y_{i,t-1} + (\beta_{int} + b_{int\,i}) + \varepsilon_{(AR)i,t} + \varepsilon_{(ME)i,t} - \rho \varepsilon_{(ME)i,t-1}, \, (t > 0).$$

The following transformation makes clear that $\varepsilon_{(ME)i,t}$ is a measurement error,

$$(Y_{i,t} - \varepsilon_{(ME)i,t}) = \rho (Y_{i,t-1} - \varepsilon_{(ME)i,t-1}) + (\beta_{int} + b_{int\,i}) + \varepsilon_{(AR)i,t}.$$
(2.4.5)

$Y_{i,t} - \varepsilon_{(ME)i,t}$ is a latent variable which would be available if there were no measurement errors. The state space representation in Sect. 6.3.1 uses this representation.

Now we show the variance covariance matrix of these error structures. Let $\mathbf{R}_{ME\,i}$ be a variance covariance matrix of a measurement error, $\varepsilon_{(ME)i,t}$, in the autoregressive form and $\mathbf{\Sigma}_{ME\,i}$ be a corresponding variance covariance matrix in the marginal form. Because the error term is independent and has a constant variance in the marginal form, $\mathbf{\Sigma}_{ME\,i} = \sigma^2_{ME} \mathbf{I}_i$. Then the autoregressive form, $\mathbf{R}_{ME\,i}$, is

Table 2.3 Variance covariance structures for autoregressive linear mixed effects models in the autoregressive form and marginal form for $T_i = 3$

Autoregressive form $\mathbf{V}_i = \mathbf{Z}_i \mathbf{G} \mathbf{Z}_i^T + \mathbf{R}_i$ $\mathbf{V}_i = (\mathbf{I}_i - \rho \mathbf{F}_i) \boldsymbol{\Sigma}_i (\mathbf{I}_i - \rho \mathbf{F}_i)^T$	Marginal (unconditional) form $\boldsymbol{\Sigma}_i = (\mathbf{I}_i - \rho \mathbf{F}_i)^{-1} \mathbf{V}_i \left\{ (\mathbf{I}_i - \rho \mathbf{F}_i)^{-1} \right\}^T$
(a) Measurement error $$\mathbf{R}_{\text{ME}\,i} = \sigma_{\text{ME}}^2 \begin{pmatrix} 1 & -\rho & 0 & 0 \\ -\rho & 1+\rho^2 & -\rho & 0 \\ 0 & -\rho & 1+\rho^2 & -\rho \\ 0 & 0 & -\rho & 1+\rho^2 \end{pmatrix}$$	(b) Measurement error $$\boldsymbol{\Sigma}_{\text{ME}\,i} = \sigma_{\text{ME}}^2 \begin{pmatrix} 1 & 0 & 0 & 0 \\ 0 & 1 & 0 & 0 \\ 0 & 0 & 1 & 0 \\ 0 & 0 & 0 & 1 \end{pmatrix}$$
(c) Non-stationary AR(1) $$\mathbf{R}_{\text{AR}\,i} = \sigma_{\text{AR}}^2 \begin{pmatrix} 0 & 0 & 0 & 0 \\ 0 & 1 & 0 & 0 \\ 0 & 0 & 1 & 0 \\ 0 & 0 & 0 & 1 \end{pmatrix}$$	(d) Non-stationary AR(1) $$\boldsymbol{\Sigma}_{\text{AR}\,i} = \sigma_{\text{AR}}^2 \begin{pmatrix} 0 & 0 & 0 & 0 \\ 0 & 1 & \rho & \rho^2 \\ 0 & \rho & 1+\rho^2 & \rho+\rho^3 \\ 0 & \rho^2 & \rho+\rho^3 & 1+\rho^2+\rho^4 \end{pmatrix}$$ When j is infinite, the (j,j)th element is $\left(1-\rho^2\right)^{-1}\sigma_{\text{AR}}^2$
(e) Stationary AR(1) $$\mathbf{R}_{\text{AR,ST}\,i} = \sigma_{\text{AR,ST}}^2 \begin{pmatrix} 1/\left(1-\rho^2\right) & 0 & 0 & 0 \\ 0 & 1 & 0 & 0 \\ 0 & 0 & 1 & 0 \\ 0 & 0 & 0 & 1 \end{pmatrix}$$	(f) Stationary AR(1) $$\boldsymbol{\Sigma}_{\text{AR,ST}\,i} = \frac{\sigma_{\text{AR,ST}}^2}{(1-\rho^2)} \begin{pmatrix} 1 & \rho & \rho^2 & \rho^3 \\ \rho & 1 & \rho & \rho^2 \\ \rho^2 & \rho & 1 & \rho \\ \rho^3 & \rho^2 & \rho & 1 \end{pmatrix}$$
(g) Random baseline $\mathbf{Z}_i \mathbf{G} \mathbf{Z}_i^T$ $$= \begin{pmatrix} 1 \\ 0 \\ 0 \\ 0 \end{pmatrix} \sigma_{\text{G0}}^2 \begin{pmatrix} 1 & 0 & 0 & 0 \end{pmatrix} = \begin{pmatrix} \sigma_{\text{G0}}^2 & 0 & 0 & 0 \\ 0 & 0 & 0 & 0 \\ 0 & 0 & 0 & 0 \\ 0 & 0 & 0 & 0 \end{pmatrix}$$	(h) Random baseline $$(\mathbf{I}_i - \rho \mathbf{F}_i)^{-1} \mathbf{Z}_i \mathbf{G} \mathbf{Z}_i^T \left\{ (\mathbf{I}_i - \rho \mathbf{F}_i)^{-1} \right\}^T$$ $$= \sigma_{\text{G0}}^2 \begin{pmatrix} 1 & \rho & \rho^2 & \rho^3 \\ \rho & \rho^2 & \rho^3 & \rho^4 \\ \rho^2 & \rho^3 & \rho^4 & \rho^5 \\ \rho^3 & \rho^4 & \rho^5 & \rho^6 \end{pmatrix}$$ When j and k are infinite, the (j,k) th element is 0
(i) Random baseline and random intercept $$\mathbf{Z}_i \mathbf{G} \mathbf{Z}_i^T = \begin{pmatrix} 1 & 0 \\ 0 & 1 \\ 0 & 1 \\ 0 & 1 \end{pmatrix} \begin{pmatrix} \sigma_{\text{G0}}^2 & \sigma_{\text{G01}} \\ \sigma_{\text{G01}} & \sigma_{\text{G1}}^2 \end{pmatrix} \begin{pmatrix} 1 & 0 & 0 & 0 \\ 0 & 1 & 1 & 1 \end{pmatrix}$$ $$= \begin{pmatrix} \sigma_{\text{G0}}^2 & \sigma_{\text{G01}} & \sigma_{\text{G01}} & \sigma_{\text{G01}} \\ \sigma_{\text{G01}} & \sigma_{\text{G1}}^2 & \sigma_{\text{G1}}^2 & \sigma_{\text{G1}}^2 \\ \sigma_{\text{G01}} & \sigma_{\text{G1}}^2 & \sigma_{\text{G1}}^2 & \sigma_{\text{G1}}^2 \\ \sigma_{\text{G01}} & \sigma_{\text{G1}}^2 & \sigma_{\text{G1}}^2 & \sigma_{\text{G1}}^2 \end{pmatrix}$$	(j) Random baseline and random intercept $$(\mathbf{I}_i - \rho \mathbf{F}_i)^{-1} \mathbf{Z}_i \mathbf{G} \mathbf{Z}_i^T \left\{ (\mathbf{I}_i - \rho \mathbf{F}_i)^{-1} \right\}^T$$ The $(j+1, k+1)$ th element is $\rho^{j+k}\sigma_{\text{G0}}^2 + \left(\rho^j + \rho^k - 2\rho^{j+k}\right)(1-\rho)^{-1}\sigma_{\text{G01}} + \left(1-\rho^j\right)\left(1-\rho^k\right)(1-\rho)^{-2}\sigma_{\text{G1}}^2$ When j and k are infinite, it is $\left(1-\rho^2\right)^{-1}\sigma_{\text{G1}}^2$

$$\mathbf{R}_{\text{ME }i} = (\mathbf{I}_i - \rho \mathbf{F}_i) \boldsymbol{\Sigma}_{\text{ME }i} (\mathbf{I}_i - \rho \mathbf{F}_i)^T. \tag{2.4.6}$$

Table 2.3a and b shows these matrices for $T_i = 3$. $\mathbf{R}_{\text{ME }i}$ is a structure similar to two-band Toeplitz in Table 1.1m except for the $(1, 1)$th element.

Let $\mathbf{R}_{\text{AR }i}$ and $\mathbf{R}_{\text{AR,ST }i}$ be variance covariance matrices of AR(1) errors of $\varepsilon_{(\text{AR})i,t}$ and $\varepsilon_{(\text{AR,ST})i,t}$ in the autoregressive form, and $\boldsymbol{\Sigma}_{\text{AR }i}$ and $\boldsymbol{\Sigma}_{\text{AR,ST }i}$ be corresponding variance covariance matrices in the marginal form. Table 2.3c–f shows these matrices. If the model has a random baseline effect $b_{\text{base }i}$, we cannot decide whether the AR(1) error is stationary based on the model fit, as mentioned in Sect. 2.4.2. The parameter σ_{AR}^2 depends on the unit time. The jth diagonal element of $\boldsymbol{\Sigma}_{\text{AR }i}$ is $\sigma_{\text{AR}}^2 (1 + \rho^2 + \cdots + \rho^{2(j-1)})$. When j is infinite, it is $(1 - \rho^2)^{-1} \sigma_{\text{AR}}^2$ and does not depend on the unit time.

2.4.2 Variance Covariance Matrix Induced by Random Effects

Next, we consider the contribution of random effects to the variance covariance matrices of the response vector. As shown in the model (2.2.1), let the baseline level, $b_{\text{base }i}(t = 0)$, and the intercept, $b_{\text{int }i}(t > 0)$, be random effects. $\mathbf{b}_i = (b_{\text{base }i}, b_{\text{int }i})^T$ is assumed to follow a bivariate normal distribution with the mean zero vector and the variance covariance matrix $\mathbf{G}$ with the variances σ_{G0}^2 and σ_{G1}^2, and covariance σ_{G01}. It is expressed as

$$\begin{pmatrix} b_{\text{base }i} \\ b_{\text{int }i} \end{pmatrix} \sim \text{MVN} \left(\begin{pmatrix} 0 \\ 0 \end{pmatrix}, \begin{pmatrix} \sigma_{\text{G0}}^2 & \sigma_{\text{G01}} \\ \sigma_{\text{G01}} & \sigma_{\text{G1}}^2 \end{pmatrix} \right). \tag{2.4.7}$$

For $T_i = 3$, the autoregressive form of the variance covariance matrix of the response vector induced by the random effects is

$$\text{Var}(\mathbf{Z}_i \mathbf{b}_i) = \mathbf{Z}_i \mathbf{G} \mathbf{Z}_i^T$$

$$= \begin{pmatrix} 1 & 0 \\ 0 & 1 \\ 0 & 1 \\ 0 & 1 \end{pmatrix} \begin{pmatrix} \sigma_{\text{G0}}^2 & \sigma_{\text{G01}} \\ \sigma_{\text{G01}} & \sigma_{\text{G1}}^2 \end{pmatrix} \begin{pmatrix} 1 & 0 & 0 & 0 \\ 0 & 1 & 1 & 1 \end{pmatrix} = \begin{pmatrix} \sigma_{\text{G0}}^2 & \sigma_{\text{G01}} & \sigma_{\text{G01}} & \sigma_{\text{G01}} \\ \sigma_{\text{G01}} & \sigma_{\text{G1}}^2 & \sigma_{\text{G1}}^2 & \sigma_{\text{G1}}^2 \\ \sigma_{\text{G01}} & \sigma_{\text{G1}}^2 & \sigma_{\text{G1}}^2 & \sigma_{\text{G1}}^2 \\ \sigma_{\text{G01}} & \sigma_{\text{G1}}^2 & \sigma_{\text{G1}}^2 & \sigma_{\text{G1}}^2 \end{pmatrix}.$$

This is a block diagonal matrix. The transformation to the marginal form is

$$\text{Var}\left((\mathbf{I}_i - \rho \mathbf{F}_i)^{-1} \mathbf{Z}_i \mathbf{b}_i \right) = (\mathbf{I}_i - \rho \mathbf{F}_i)^{-1} \mathbf{Z}_i \mathbf{G} \mathbf{Z}_i^T \left\{ (\mathbf{I}_i - \rho \mathbf{F}_i)^{-1} \right\}^T$$

$$
= \begin{pmatrix} 1 & 0 & 0 & 0 \\ \rho & 1 & 0 & 0 \\ \rho^2 & \rho & 1 & 0 \\ \rho^3 & \rho^2 & \rho & 1 \end{pmatrix} \begin{pmatrix} \sigma_{G0}^2 & \sigma_{G01} & \sigma_{G01} & \sigma_{G01} \\ \sigma_{G01} & \sigma_{G1}^2 & \sigma_{G1}^2 & \sigma_{G1}^2 \\ \sigma_{G01} & \sigma_{G1}^2 & \sigma_{G1}^2 & \sigma_{G1}^2 \\ \sigma_{G01} & \sigma_{G1}^2 & \sigma_{G1}^2 & \sigma_{G1}^2 \end{pmatrix} \begin{pmatrix} 1 & \rho & \rho^2 & \rho^3 \\ 0 & 1 & \rho & \rho^2 \\ 0 & 0 & 1 & \rho \\ 0 & 0 & 0 & 1 \end{pmatrix}.
$$

The $(j + 1, k + 1)$th element of this matrix is

$$
\rho^{j+k}\sigma_{G0}^2 + \left(\frac{\rho^j + \rho^k - 2\rho^{j+k}}{1 - \rho} \right)\sigma_{G01} + \frac{(1 - \rho^j)(1 - \rho^k)\sigma_{G1}^2}{(1 - \rho)^2}. \tag{2.4.8}
$$

When both j and k are infinite, the (j, k)th element is $(1 - \rho^2)^{-1}\sigma_{G1}^2$; this value represents the inter-individual variance of the random asymptotes. Table 2.3i and j shows these matrices.

Next, we consider a case with three random effects. As shown in the model (2.2.8), let the baseline level, $b_{\text{base }i}(t = 0)$, the intercept, $b_{\text{int }i}(t > 0)$, and a covariate effect, $b_{\text{cov }i}$, be random effects. $\mathbf{b}_i = (b_{\text{base }i}, b_{\text{int }i}, b_{\text{cov }i})^T$ is assumed to follow the trivariate normal distribution,

$$
\begin{pmatrix} b_{\text{base }i} \\ b_{\text{int }i} \\ b_{\text{cov }i} \end{pmatrix} \sim \text{MVN}\left(\begin{pmatrix} 0 \\ 0 \\ 0 \end{pmatrix}, \begin{pmatrix} \sigma_{Gb}^2 & \sigma_{Gbi} & \sigma_{Gbc} \\ \sigma_{Gbi} & \sigma_{Gi}^2 & \sigma_{Gic} \\ \sigma_{Gbc} & \sigma_{Gic} & \sigma_{Gc}^2 \end{pmatrix} \right). \tag{2.4.9}
$$

For $T_i = 3$, the variance covariance matrix of the response vector induced by the random effects is

$$
\text{Var}(\mathbf{Z}_i\mathbf{b}_i) = \mathbf{Z}_i\mathbf{G}\mathbf{Z}_i^T = \begin{pmatrix} 1 & 0 & 0 \\ 0 & 1 & x_{i,1} \\ 0 & 1 & x_{i,2} \\ 0 & 1 & x_{i,3} \end{pmatrix} \begin{pmatrix} \sigma_{Gb}^2 & \sigma_{Gbi} & \sigma_{Gbc} \\ \sigma_{Gbi} & \sigma_{Gi}^2 & \sigma_{Gic} \\ \sigma_{Gbc} & \sigma_{Gic} & \sigma_{Gc}^2 \end{pmatrix} \begin{pmatrix} 1 & 0 & 0 & 0 \\ 0 & 1 & 1 & 1 \\ 0 & x_{i,1} & x_{i,2} & x_{i,3} \end{pmatrix}.
$$

Table 2.3g shows the autoregressive form of the variance covariance structure induced by a random baseline, $b_{\text{base }i}$, for $T_i = 3$. σ_{G0}^2 is the $(1, 1)$th element of $\mathbf{V}_i$, and the difference in the assumptions between $\mathbf{R}_{\text{AR }i}$ and $\mathbf{R}_{\text{AR,ST }i}$ is also on the $(1, 1)$th element in Table 2.3c and e. Therefore, if the model has a random baseline effect, we cannot decide whether the AR(1) error is stationary or not based on the model fit. Table 2.3h shows the corresponding marginal form.

2.4.3 Variance Covariance Matrix Induced by Random Effects and Random Errors

We will now consider the variance covariance matrix of the response vector when the random effects, $b_{\text{base }i}$ and $b_{\text{int }i}$, an AR(1) error, $\varepsilon_{(AR)i,t}$, and a measurement error, $\varepsilon_{(ME)i,t}$, are assumed simultaneously. First, we assume a non-stationary AR(1) error with $\varepsilon_{(AR)i,0} = 0$. For $T_i = 3$, Table 2.4a shows the autoregressive form $V_i = Z_i G Z_i^T + R_i$. Table 2.4b shows the corresponding marginal form Σ_i, where A is a 4×4 matrix, and the $(j + 1, k + 1)$ th element is (2.4.8). When j and k are infinite, the random effects and the measurement errors produce a compound symmetry (CS) structure with the diagonal elements being $(1 - \rho^2)^{-1}\sigma_{G1}^2 + \sigma_{ME}^2$ and the nondiagonal elements being $(1 - \rho^2)^{-1}\sigma_{G1}^2$. Next, we assume a stationary AR(1) error $\varepsilon_{(AR,ST)i,t}$ instead of $\varepsilon_{(AR)i,t}$. Table 2.4c shows the autoregressive form of the variance covariance structure of the response vector V_i. Table 2.4d shows the corresponding marginal form Σ_i.

Table 2.4 Examples of variance covariance matrices of the response vector induced by random effects and random errors in the autoregressive form and marginal form for $T_i = 3$

Representation, assumption on AR(1) error

(a) Autoregressive form V_i, non-stationary AR(1)

$$Z_i G Z_i^T + R_{ARi} + R_{MEi} = \begin{pmatrix} \sigma_{G0}^2 & \sigma_{G01} & \sigma_{G01} & \sigma_{G01} \\ \sigma_{G01} & \sigma_{G1}^2 & \sigma_{G1}^2 & \sigma_{G1}^2 \\ \sigma_{G01} & \sigma_{G1}^2 & \sigma_{G1}^2 & \sigma_{G1}^2 \\ \sigma_{G01} & \sigma_{G1}^2 & \sigma_{G1}^2 & \sigma_{G1}^2 \end{pmatrix} + \sigma_{AR}^2 \begin{pmatrix} 0 & 0 & 0 & 0 \\ 0 & 1 & 0 & 0 \\ 0 & 0 & 1 & 0 \\ 0 & 0 & 0 & 1 \end{pmatrix} + \sigma_{ME}^2 \begin{pmatrix} 1 & -\rho & 0 & 0 \\ -\rho & 1+\rho^2 & -\rho & 0 \\ 0 & -\rho & 1+\rho^2 & -\rho \\ 0 & 0 & -\rho & 1+\rho^2 \end{pmatrix}$$

(b)[a] Marginal (unconditional) form Σ_i, non-stationary AR(1)

$(I_i - \rho F_i)^{-1}(Z_i G Z_i^T)\{(I_i - \rho F_i)^{-1}\}^T + \Sigma_{ARi} + \Sigma_{MEi}$

$$= A + \sigma_{AR}^2 \begin{pmatrix} 0 & 0 & 0 & 0 \\ 0 & 1 & \rho & \rho^2 \\ 0 & \rho & 1+\rho^2 & \rho+\rho^3 \\ 0 & \rho^2 & \rho+\rho^3 & 1+\rho^2+\rho^4 \end{pmatrix} + \sigma_{ME}^2 \begin{pmatrix} 1 & 0 & 0 & 0 \\ 0 & 1 & 0 & 0 \\ 0 & 0 & 1 & 0 \\ 0 & 0 & 0 & 1 \end{pmatrix}$$

(c) Autoregressive form V_i, stationary AR(1)

$$Z_i G Z_i^T + R_{AR,STi} + R_{MEi} = \begin{pmatrix} \sigma_{G0}^2 & \sigma_{G01} & \sigma_{G01} & \sigma_{G01} \\ \sigma_{G01} & \sigma_{G1}^2 & \sigma_{G1}^2 & \sigma_{G1}^2 \\ \sigma_{G01} & \sigma_{G1}^2 & \sigma_{G1}^2 & \sigma_{G1}^2 \\ \sigma_{G01} & \sigma_{G1}^2 & \sigma_{G1}^2 & \sigma_{G1}^2 \end{pmatrix} + \sigma_{AR,ST}^2 \begin{pmatrix} \frac{1}{1-\rho^2} & 0 & 0 & 0 \\ 0 & 1 & 0 & 0 \\ 0 & 0 & 1 & 0 \\ 0 & 0 & 0 & 1 \end{pmatrix} + \sigma_{ME}^2 \begin{pmatrix} 1 & -\rho & 0 & 0 \\ -\rho & 1+\rho^2 & -\rho & 0 \\ 0 & -\rho & 1+\rho^2 & -\rho \\ 0 & 0 & -\rho & 1+\rho^2 \end{pmatrix}$$

(d)[a] Marginal (unconditional) form Σ_i, stationary AR(1)

$$(I_i - \rho F_i)^{-1}(Z_i G Z_i^T)\{(I_i - \rho F_i)^{-1}\}^T + \Sigma_{AR,STi} + \Sigma_{MEi} = A + \frac{\sigma_{AR,ST}^2}{(1-\rho^2)} \begin{pmatrix} 1 & \rho & \rho^2 & \rho^3 \\ \rho & 1 & \rho & \rho^2 \\ \rho^2 & \rho & 1 & \rho \\ \rho^3 & \rho^2 & \rho & 1 \end{pmatrix} + \sigma_{ME}^2 \begin{pmatrix} 1 & 0 & 0 & 0 \\ 0 & 1 & 0 & 0 \\ 0 & 0 & 1 & 0 \\ 0 & 0 & 0 & 1 \end{pmatrix}$$

[a] The $(j + 1, k + 1)$ th element of A is $\rho^{j+k}\sigma_{G0}^2 + (\rho^j + \rho^k - 2\rho^{j+k})(1 - \rho)^{-1}\sigma_{G01} + (1 - \rho^j)(1 - \rho^k)(1 - \rho)^{-2}\sigma_{G1}^2$

2.4.4 Variance Covariance Matrix for Asymptotes

In Sect. 2.3.2, fixed effects and random effects parameters, $\boldsymbol{\beta}$ and $\mathbf{b}_i$, were multiplied by matrices, $\mathbf{M}_x$ and $\mathbf{M}_z$, and transformed to the parameters for the asymptotes, $\boldsymbol{\beta}^*$ and $\mathbf{b}_i^*$. In this section, we consider the variance covariance matrix of $\mathbf{b}_i^*$.

First, we consider the case of two random effects, $\mathbf{b}_i = (b_{\text{base } i}, b_{\text{int } i})^T$, as shown in the model (2.2.1). Let $\mathbf{M}_z$ be a diagonal matrix with diagonal elements $\left(1, (1-\rho)^{-1}\right)$. Then, $\mathbf{b}_i$ is transformed to $\mathbf{b}_i^* = \mathbf{M}_z \mathbf{b}_i$,

$$
\mathbf{b}_i^* = \begin{pmatrix} b_{\text{base } i} \\ b_{\text{int } i}^* \end{pmatrix} = \begin{pmatrix} 1 & 0 \\ 0 & (1-\rho)^{-1} \end{pmatrix} \begin{pmatrix} b_{\text{base } i} \\ b_{\text{int } i} \end{pmatrix} = \begin{pmatrix} b_{\text{base } i} \\ (1-\rho)^{-1} b_{\text{int } i} \end{pmatrix}. \tag{2.4.10}
$$

Here, * (asterisk) shows the parameters for the asymptote. $\mathbf{b}_i$ and $\mathbf{b}_i^*$ follow multivariate normal distributions,

$$
\mathbf{b}_i \sim \text{MVN}(\mathbf{0}, \mathbf{G}), \tag{2.4.11}
$$

$$
\mathbf{b}_i^* \sim \text{MVN}\left(\mathbf{0}, \mathbf{M}_z \mathbf{G} \mathbf{M}_z^T\right). \tag{2.4.12}
$$

The variance covariance matrix of $\mathbf{b}_i^*$ is

$$
\text{Var}(\mathbf{b}_i^*) = \text{Var}(\mathbf{M}_z \mathbf{b}_i) = \mathbf{M}_z \mathbf{G} \mathbf{M}_z^T = \begin{pmatrix} 1 & 0 \\ 0 & (1-\rho)^{-1} \end{pmatrix} \begin{pmatrix} \sigma_{G0}^2 & \sigma_{G01} \\ \sigma_{G01} & \sigma_{G1}^2 \end{pmatrix} \begin{pmatrix} 1 & 0 \\ 0 & (1-\rho)^{-1} \end{pmatrix}
$$

$$
= \begin{pmatrix} \sigma_{G0}^2 & (1-\rho)^{-1} \sigma_{G01} \\ (1-\rho)^{-1} \sigma_{G01} & (1-\rho)^{-2} \sigma_{G1}^2 \end{pmatrix}. \tag{2.4.13}
$$

$(1-\rho)^{-1} \sigma_{G01}$ is the covariance of the random baseline and random asymptote. The correlation is

$$
\text{Corr}\left(b_{\text{base } i}, b_{\text{int } i}^*\right) = \frac{\sigma_{G01}}{(1-\rho)\sigma_{G0}\sigma_{G1}}. \tag{2.4.14}
$$

Next, we consider the case of three random effects $\mathbf{b}_i = (b_{\text{base } i}, b_{\text{int } i}, b_{\text{cov } i})^T$ as shown in the model (2.2.8). Let $\mathbf{M}_z$ be a diagonal matrix with diagonal elements $\left(1, (1-\rho)^{-1}, (1-\rho)^{-1}\right)$. $\mathbf{b}_i$ is transformed to $\mathbf{b}_i^* = \mathbf{M}_z \mathbf{b}_i$,

$$
\mathbf{b}_i^* = \begin{pmatrix} b_{\text{base } i} \\ b_{\text{int } i}^* \\ b_{\text{cov } i}^* \end{pmatrix} = \begin{pmatrix} 1 & 0 & 0 \\ 0 & (1-\rho)^{-1} & 0 \\ 0 & 0 & (1-\rho)^{-1} \end{pmatrix} \begin{pmatrix} b_{\text{base } i} \\ b_{\text{int } i} \\ b_{\text{cov } i} \end{pmatrix} = \begin{pmatrix} b_{\text{base } i} \\ (1-\rho)^{-1} b_{\text{int } i} \\ (1-\rho)^{-1} b_{\text{cov } i} \end{pmatrix} \tag{2.4.15}
$$

The variance covariance matrix of $\mathbf{b}_i^*$, $\mathbf{M}_z\mathbf{G}\mathbf{M}_z^T$ is

$$\mathbf{M}_z\mathbf{G}\mathbf{M}_z^T = \begin{pmatrix} 1 & 0 & 0 \\ 0 & (1-\rho)^{-1} & 0 \\ 0 & 0 & (1-\rho)^{-1} \end{pmatrix} \begin{pmatrix} \sigma_{Gb}^2 & \sigma_{Gbi} & \sigma_{Gbc} \\ \sigma_{Gbi} & \sigma_{Gi}^2 & \sigma_{Gic} \\ \sigma_{Gbc} & \sigma_{Gic} & \sigma_{Gc}^2 \end{pmatrix} \begin{pmatrix} 1 & 0 & 0 \\ 0 & (1-\rho)^{-1} & 0 \\ 0 & 0 & (1-\rho)^{-1} \end{pmatrix}$$

$$= \begin{pmatrix} \sigma_{Gb}^2 & (1-\rho)^{-1}\sigma_{Gbi} & (1-\rho)^{-1}\sigma_{Gbc} \\ (1-\rho)^{-1}\sigma_{Gbi} & (1-\rho)^{-2}\sigma_{Gi}^2 & (1-\rho)^{-2}\sigma_{Gic} \\ (1-\rho)^{-1}\sigma_{Gbc} & (1-\rho)^{-2}\sigma_{Gic} & (1-\rho)^{-2}\sigma_{Gc}^2 \end{pmatrix}. \qquad (2.4.16)$$

2.5 Estimation in Autoregressive Linear Mixed Effects Models

Similar to the linear mixed effects models in Chap. 1, maximum likelihood (ML) methods are used for estimation in autoregressive linear mixed effects models. There are several ways to calculate likelihood. Sections 2.5.1 and 2.5.2 show likelihood of marginal and autoregressive forms, respectively. Likelihood of the marginal form can be also calculated using the Kalman filter in Chap. 6. Section 2.5.3 explains the method to obtain indirectly the ML estimates from the autoregressive form using estimation methods of linear mixed effects models.

2.5.1 Likelihood of Marginal Form

The marginal form of autoregressive linear mixed effects models is

$$\mathbf{Y}_i = (\mathbf{I}_i - \rho\mathbf{F}_i)^{-1}(\mathbf{X}_i\boldsymbol{\beta} + \mathbf{Z}_i\mathbf{b}_i + \boldsymbol{\varepsilon}_i). \qquad (2.5.1)$$

The response vector follows a multivariate normal distribution with the mean,

$$E(\mathbf{Y}_i) = (\mathbf{I}_i - \rho\mathbf{F}_i)^{-1}\mathbf{X}_i\boldsymbol{\beta}, \qquad (2.5.2)$$

and the variance covariance matrix,

$$\boldsymbol{\Sigma}_i = (\mathbf{I}_i - \rho\mathbf{F}_i)^{-1}\mathbf{V}_i\left\{(\mathbf{I}_i - \rho\mathbf{F}_i)^{-1}\right\}^T. \qquad (2.5.3)$$

Then, the marginal (unconditional) form of –2 log-likelihood ($-2ll$) is

$$-2ll_{\mathrm{ML}} = \sum_{i=1}^{N} \left[n_i \log(2\pi) + \log|\mathbf{\Sigma}_i| + \left\{ \mathbf{Y}_i - (\mathbf{I}_i - \rho\mathbf{F}_i)^{-1}\mathbf{X}_i\boldsymbol{\beta} \right\}^T \mathbf{\Sigma}_i^{-1} \left\{ \mathbf{Y}_i - (\mathbf{I}_i - \rho\mathbf{F}_i)^{-1}\mathbf{X}_i\boldsymbol{\beta} \right\} \right].$$

(2.5.4)

When the variance covariance parameters and the autoregressive parameter, ρ, are known, the ML estimators (MLEs) of the fixed effects are given by

$$\hat{\boldsymbol{\beta}} = \left[\sum_{i=1}^{N} \left\{ (\mathbf{I}_i - \rho\mathbf{F}_i)^{-1}\mathbf{X}_i \right\}^T \mathbf{\Sigma}_i^{-1}(\mathbf{I}_i - \rho\mathbf{F}_i)^{-1}\mathbf{X}_i \right]^{-1} \sum_{i=1}^{N} \left\{ (\mathbf{I}_i - \rho\mathbf{F}_i)^{-1}\mathbf{X}_i \right\}^T \mathbf{\Sigma}_i^{-1}\mathbf{Y}_i.$$

(2.5.5)

To obtain the ML estimates of the variance covariance parameters and ρ, we substitute $\hat{\boldsymbol{\beta}}$ for $\boldsymbol{\beta}$ in $-2ll_{\mathrm{ML}}$ (2.5.4). This concentrated $-2ll_{\mathrm{ML\ CONC}}$ is expressed as

$$-2ll_{\mathrm{ML\ CONC}} = \sum_{i=1}^{N} n_i \log(2\pi) + \sum_{i=1}^{N} \log|\mathbf{\Sigma}_i| + \sum_{i=1}^{N} \mathbf{Y}_i^T \mathbf{\Sigma}_i^{-1}\mathbf{Y}_i$$

$$- \left\{ \sum_{i=1}^{N} \mathbf{Y}_i^T \mathbf{\Sigma}_i^{-1}(\mathbf{I}_i - \rho\mathbf{F}_i)^{-1}\mathbf{X}_i \right\} \hat{\boldsymbol{\beta}}.$$

(2.5.6)

We can calculate $-2ll_{\mathrm{ML\ CONC}}$ based on (2.5.6), and we can also use the modified Kalman filter presented in Chap. 6. We minimize $-2ll_{\mathrm{ML\ CONC}}$ by optimization methods. If some elements of $\mathbf{Y}_i$ are intermittently missing but the corresponding elements of $\mathbf{X}_i$ are known, we can use this equation deleting the missing part and the corresponding parts of $(\mathbf{I}_i - \rho\mathbf{F}_i)^{-1}\mathbf{X}_i$ and $\mathbf{\Sigma}_i$.

The standard errors of the ML estimates are derived from the Hessian of the log-likelihood. To obtain the standard errors, the fixed effects parameters are included in the log-likelihood. It is necessary to parameterize the intended parameters, such as the asymptote, directly. The Hessian at the ML estimates can be obtained by numerical methods. The random effects are predicted by

$$\hat{\mathbf{b}}_i = \widehat{\mathbf{G}}\left\{ (\mathbf{I}_i - \hat{\rho}\mathbf{F}_i)^{-1}\mathbf{Z}_i \right\}^T \widehat{\mathbf{\Sigma}}_i^{-1} \left\{ \mathbf{Y}_i - (\mathbf{I}_i - \hat{\rho}\mathbf{F}_i)^{-1}\mathbf{X}_i\hat{\boldsymbol{\beta}} \right\}.$$

(2.5.7)

2.5.2   Likelihood of Autoregressive Form

The autoregressive form of autoregressive linear mixed effects models given the previous response is

$$\mathbf{Y}_i = \rho\mathbf{F}_i\mathbf{Y}_i + \mathbf{X}_i\boldsymbol{\beta} + \mathbf{Z}_i\mathbf{b}_i + \boldsymbol{\varepsilon}_i.$$

(2.5.8)

The response vector follows a multivariate normal distribution with the mean $\rho\mathbf{F}_i\mathbf{Y}_i + \mathbf{X}_i\boldsymbol{\beta}$ and the variance covariance matrix $\mathbf{V}_i$. Then, the autoregressive form of $-2ll$ is

$$-2ll_{\text{ML A}} = \sum_{i=1}^{N} \left\{ n_i \log(2\pi) + \log|\mathbf{V}_i| + (\mathbf{Y}_i - \rho \mathbf{F}_i \mathbf{Y}_i - \mathbf{X}_i \boldsymbol{\beta})^T \mathbf{V}_i^{-1} (\mathbf{Y}_i - \rho \mathbf{F}_i \mathbf{Y}_i - \mathbf{X}_i \boldsymbol{\beta}) \right\}.$$

$$(2.5.9)$$

where $\log|\mathbf{V}_i|$ is equal to $\log|\boldsymbol{\Sigma}_i|$. When there are no intermittently missing response values, $-2ll_{\text{ML}}$ (2.5.4) and $-2ll_{\text{ML A}}$ give the same value, and we can obtain the MLEs using either equation. However, when elements are intermittently missing, we cannot calculate the autoregressive form (2.5.9). When the variance covariance parameters and the autoregressive parameter, ρ, are known, the MLEs of the fixed effects are given by

$$\hat{\boldsymbol{\beta}} = \left(\sum_{i=1}^{N} \mathbf{X}_i^T \mathbf{V}_i^{-1} \mathbf{X}_i \right)^{-1} \sum_{i=1}^{N} \mathbf{X}_i^T \mathbf{V}_i^{-1} (\mathbf{I}_i - \rho \mathbf{F}_i) \mathbf{Y}_i. \qquad (2.5.10)$$

To obtain the ML estimates of the variance covariance parameters and ρ, we substitute $\hat{\boldsymbol{\beta}}$ for $\boldsymbol{\beta}$ in $-2ll_{\text{ML A}}$ (2.5.9), and minimize the concentrated $-2ll_{\text{ML A CONC}}$ by optimization methods.

2.5.3   Indirect Methods Using Linear Mixed Effects Models

The autoregressive form (2.5.8) can be expressed by

$$\mathbf{Y}_i = \mathbf{X}_i^\# \boldsymbol{\beta}^\# + \mathbf{Z}_i \mathbf{b}_i + \boldsymbol{\varepsilon}_i, \qquad (2.5.11)$$

where $\boldsymbol{\beta}^\# = \left(\boldsymbol{\beta}^T, \rho \right)^T$ and $\mathbf{X}_i^\# = (\mathbf{X}_i, \mathbf{F}_i \mathbf{Y}_i)$. The $-2ll$ of the autoregressive form (2.5.9) is expressed by

$$-2ll_{\text{ML L}} = \sum_{i=1}^{N} \left\{ n_i \log(2\pi) + \log|\mathbf{V}_i| + \left(\mathbf{Y}_i - \mathbf{X}_i^\# \boldsymbol{\beta}^\# \right)^T \mathbf{V}_i^{-1} \left(\mathbf{Y}_i - \mathbf{X}_i^\# \boldsymbol{\beta}^\# \right) \right\}.$$

$$(2.5.12)$$

These are the same forms with linear mixed effects models (1.2.1) and (1.5.7) in Chap. 1, but $\boldsymbol{\beta}^\#$ and $\mathbf{V}_i$ include the unknown parameter ρ. However, in some covariance structures such as the variance covariance structure in Table 2.4a and c, a known $\mathbf{V}_i$ does not specify the variance covariance parameters and ρ simultaneously. In these cases, we can use an indirect method based on (2.5.12). When $\mathbf{V}_i$ is known, the fixed effects $\boldsymbol{\beta}$ and ρ are estimated by

$$\hat{\boldsymbol{\beta}}^\# = \left(\sum_{i=1}^{N} \mathbf{X}_i^{\#T} \mathbf{V}_i^{-1} \mathbf{X}_i^\# \right)^{-1} \sum_{i=1}^{N} \mathbf{X}_i^{\#T} \mathbf{V}_i^{-1} \mathbf{Y}_i. \qquad (2.5.13)$$

To obtain the estimates of the variance covariance parameters, we substitute $\hat{\beta}^{\#}$ for $\beta^{\#}$ in $-2ll_{\mathrm{ML\,L}}$ (2.5.12) and minimize it using optimization methods.

In the variance covariance structure in Table 2.4a, the random baseline and the random intercept are correlated and there are a measurement error and an autoregressive error. Although there are six parameters, ρ, σ_{G0}^2, σ_{G1}^2, σ_{G01}, σ_{ME}^2, and σ_{AR}^2, there are five distinct elements in $\mathbf{V}_i$: $\sigma_{G0}^2 + \sigma_{ME}^2$ for (1, 1)th, $\sigma_{G0}^2 + \sigma_{AR}^2 + (1 + \rho^2)\sigma_{ME}^2$ for (j, j)th $(j > 1)$, $\sigma_{G01} - \rho\sigma_{ME}^2$ for $(j, j + 1)$ and $(j, j - 1)$th, σ_{G01} for $(1, j)$ and $(j, 1)$th $(j > 2)$, and σ_{G1}^2 otherwise. In this case, we can use the following modified estimation method. First, we reparameterize $\mathbf{V}_i$ as

$$
\mathbf{V}_i = \begin{pmatrix} \sigma_{G0}^{2\#} & \sigma_{G01} & \sigma_{G01} & \sigma_{G01} \\ \sigma_{G01} & \sigma_{G1}^2 & \sigma_{G1}^2 & \sigma_{G1}^2 \\ \sigma_{G01} & \sigma_{G1}^2 & \sigma_{G1}^2 & \sigma_{G1}^2 \\ \sigma_{G01} & \sigma_{G1}^2 & \sigma_{G1}^2 & \sigma_{G1}^2 \end{pmatrix} + \begin{pmatrix} R_{\mathrm{diag}} & R_{\mathrm{sub}} & 0 & 0 \\ R_{\mathrm{sub}} & R_{\mathrm{diag}} & R_{\mathrm{sub}} & 0 \\ 0 & R_{\mathrm{sub}} & R_{\mathrm{diag}} & R_{\mathrm{sub}} \\ 0 & 0 & R_{\mathrm{sub}} & R_{\mathrm{diag}} \end{pmatrix},
\tag{2.5.14}
$$

where σ_{G01} and σ_{G1}^2 are common between the structures in Table 2.4a and (2.5.14) and $R_{\mathrm{diag}} = \sigma_{AR}^2 + (1 + \rho^2)\sigma_{ME}^2$, $R_{\mathrm{sub}} = -\rho\sigma_{ME}^2$, and $\sigma_{G0}^{2\#} = -R_{\mathrm{diag}} + \sigma_{G0}^2 + \sigma_{ME}^2 = \sigma_{G0}^2 - \sigma_{AR}^2 - \rho^2\sigma_{ME}^2$. Then, σ_{ME}^2, σ_{AR}^2, and σ_{G0}^2 are

$$
\sigma_{ME}^2 = -R_{\mathrm{sub}}/\rho,
\tag{2.5.15}
$$

$$
\sigma_{AR}^2 = R_{\mathrm{diag}} - (1 + \rho^2)\sigma_{ME}^2,
\tag{2.5.16}
$$

$$
\sigma_{G0}^2 = \sigma_{G0}^{2\#} + \sigma_{AR}^2 + \rho^2\sigma_{ME}^2.
\tag{2.5.17}
$$

If $\sigma_{AR,ST}^2$ is assumed instead of σ_{AR}^2 as the structure in Table 2.4c, the (1, 1)th element in $\mathbf{V}_i$ is $\sigma_{G0}^2 + (1 - \rho^2)^{-1}\sigma_{AR,ST}^2 + \sigma_{ME}^2$. $\sigma_{AR,ST}^2$ is given by (2.5.16) and σ_{G0}^2 is

$$
\sigma_{G0}^2 = \sigma_{G0}^{2\#} - \rho^2(1 - \rho^2)^{-1}\sigma_{AR,ST}^2 + \rho^2\sigma_{ME}^2.
\tag{2.5.18}
$$

This method uses an estimation method for linear mixed effects models. It is useful in practice, because standard software for longitudinal data analysis can be used for the estimation if it supports these variance covariance structures. The ML estimation of the SAS MIXED procedure, for example, is used. It uses a ridge-stabilized Newton–Raphson algorithm for optimization. However, this method cannot be used when the previous response or the covariates are missing. Furthermore, this method does not provide the standard errors of the parameters directly. This method may be useful to find initial values of parameters before optimization for Sects. 2.5.1 and 2.5.2.

Here, we provide the SAS code of the MIXED procedure to obtain the ML estimates of the following model:

$$\begin{cases} Y_{i,0} = \beta_{g,\text{base}} + b_{\text{base }i} + \varepsilon_{(AR,ST)i,0} + \varepsilon_{(ME)i,0} \\ Y_{i,t} = \rho Y_{i,t-1} + \beta_{g,\text{int}} + b_{\text{int }i} + \varepsilon_{(AR,ST)i,t} + \varepsilon_{(ME)i,t} - \rho\varepsilon_{(ME)i,t-1}, \; (t>0) \end{cases},$$

$$(2.5.19)$$

where $g = 0$ for group A and $g = 1$ for group B. $\beta_{g,\text{base}}$ and $\beta_{g,\text{int}}$ with the subscript g represent the parameters of the baseline and intercept, $\beta_{0,\text{base}}$, $\beta_{1,\text{base}}$, $\beta_{0,\text{int}}$, and $\beta_{1,\text{int}}$ for the two groups. Although we use a different notation and parameters from (2.2.6), the assumed mean structure is the same. The dataset for the first subject with $g = 1$ and $\mathbf{Y}_1 = (93, 93, 49, 40, 46, 62)^T$ is also provided below. Note that this code indirectly provides the ML estimates and does not provide standard errors. This cord can be used for more than two groups, because the class statement produces indicator variables for the specified qualitative variable. This code is used for either variance covariance structure in Table 2.4a and c.

```
id group time y yt1 t1 t2
1 1 0 93 0   1 0
1 1 1 93 93 0 1
1 1 2 49 93 0 1
1 1 3 40 49 0 1
1 1 4 46 40 0 1
1 1 5 62 46 0 1
```

```
proc mixed method=ml;
        class group id time;
        model y=group*t2 group*t1 yt1 /s noint;
        random t1 t2 / type=un sub=id;
        repeated time / type=toep(2) sub=id;
run;
```

2.6 Models with Autoregressive Error Terms

In this section, we consider the difference between autoregressive models of the response itself (2.1.1) and linear models with an AR(1) error (2.1.2). In the autoregressive models of the response itself, the baseline response, $Y_{i,0}$, is not necessarily modeled explicitly. However, the changes in the response over time depend on the assumption of the baseline response as shown in Fig. 2.3a. With a general model for the baseline, $\mathbf{X}_{i,0}\boldsymbol{\beta}_{\text{base}} + \varepsilon_{i,0}$, the marginal form of the model (2.1.1) is

$$\begin{cases} Y_{i,0} = \mathbf{X}_{i,0}\boldsymbol{\beta}_{\text{base}} + \varepsilon_{i,0} \\ Y_{i,t} = \rho^t \mathbf{X}_{i,0}\boldsymbol{\beta}_{\text{base}} + \sum_{l=1}^{t} \rho^{t-l}\mathbf{X}_{i,l}\boldsymbol{\beta} + \sum_{l=0}^{t} \rho^{t-l}\varepsilon_{i,l}, \; (t>0) \end{cases}.$$

$$(2.6.1)$$

The marginal form of the model with an AR(1) error (2.1.2) is

$$
\begin{cases}
Y_{i,t} = \mathbf{X}_{i,t}\boldsymbol{\beta}_e + \varepsilon_{e\,i,t} \\
\varepsilon_{e\,i,t} = \displaystyle\sum_{l=0}^{t} \rho^{t-l}\eta_{i,l}
\end{cases},
$$
(2.6.2)

where $\varepsilon_{e\,i,0} = \eta_{i,0}$. For a stationary process, we assume $\eta_{i,t} \sim \mathrm{N}\big(0, \sigma_\eta^2\big)(t > 0)$ and $\varepsilon_{e\,i,0} \sim \mathrm{N}\big(0, \sigma_{\varepsilon_e}^2\big)$ with $\sigma_{\varepsilon_e}^2 = \big(1 - \rho^2\big)^{-1}\sigma_\eta^2$, then $\varepsilon_{e\,i,t} \sim \mathrm{N}\big(0, \sigma_{\varepsilon_e}^2\big)$. In this model, the baseline response is explicitly modeled. If we assume $\varepsilon_{i,0} \sim \mathrm{N}\big(0, \big(1 - \rho^2\big)^{-1}\sigma_\varepsilon^2\big)$, both error structures are the same stationary AR(1) with the relationships $\varepsilon_{i,t} = \eta_{i,t}$ and $\sigma_\varepsilon^2 = \sigma_\eta^2$.

When there are time-dependent covariates, the mean structures of the two models usually differ. The current response depends on both the current covariates and the past covariate history in the model (2.6.1), but it depends only on the current covariates in the model (2.6.2).

When there are no time-dependent covariates with $\mathbf{X}_{i,j} = \mathbf{X}_{i,k}$, the mean level of the ith subject is constant in (2.6.2). On the other hand, the model (2.6.1) shows the changes to the asymptote, $\mathbf{X}_{i,t}(1 - \rho)^{-1}\boldsymbol{\beta}$, from the baseline, $\mathbf{X}_{i,0}\boldsymbol{\beta}_{\text{base}}$, and it is not a linear model. Under the constraint that the baseline equals the asymptote as $\boldsymbol{\beta}_{\text{base}} = (1 - \rho)^{-1}\boldsymbol{\beta}$, (2.6.1) is

$$
Y_{i,t} = \mathbf{X}_{i,t}(1 - \rho)^{-1}\boldsymbol{\beta} + \sum_{l=0}^{t} \rho^{t-l}\varepsilon_{i,l}.
$$
(2.6.3)

In this case, the mean level is constant, and the two models are the same with the relationship $(1 - \rho)^{-1}\boldsymbol{\beta} = \boldsymbol{\beta}_e$. However, without the constraint, the two models show different response profiles over time. We assume a separate model for the baseline response, apart from the later responses, throughout this book.

The autoregressive form of the model (2.1.2) is

$$
\begin{cases}
Y_{i,0} = \mathbf{X}_{i,0}\boldsymbol{\beta}_e + \varepsilon_{e\,i,0} \\
Y_{i,t} = \mathbf{X}_{i,t}\boldsymbol{\beta}_e + \rho\big(Y_{i,t-1} - \mathbf{X}_{i,t-1}\boldsymbol{\beta}_e\big) + \eta_{i,t} \\
\quad\ = \rho Y_{i,t-1} + \big(\mathbf{X}_{i,t} - \rho\mathbf{X}_{i,t-1}\big)\boldsymbol{\beta}_e + \eta_{i,t}, \ (t > 0)
\end{cases}.
$$
(2.6.4)

When there are no time-dependent covariates with $\mathbf{X}_{i,j} = \mathbf{X}_{i,k}$, (2.6.4) is

$$
\begin{cases}
Y_{i,0} = \mathbf{X}_{i,0}\boldsymbol{\beta}_e + \varepsilon_{e\,i,0} \\
Y_{i,t} = \rho Y_{i,t-1} + \mathbf{X}_{i,t}(1 - \rho)\boldsymbol{\beta}_e + \eta_{i,t}, \ (t > 0)
\end{cases}.
$$
(2.6.5)

References

Anderson TW, Hsiao C (1982) Formulation and estimation of dynamic models using panel data. J Econometrics 18:47–82

Diggle PJ, Heagerty P, Liang KY, Zeger SL (2002) Analysis of longitudinal data, 2nd edn. Oxford University Press

Fitzmaurice GM, Laird NM, Ware JH (2011) Applied longitudinal analysis, 2nd edn. Wiley

Funatogawa I, Funatogawa T (2012a) An autoregressive linear mixed effects model for the analysis of unequally spaced longitudinal data with dose-modification. Stat Med 31:589–599

Funatogawa I, Funatogawa T (2012b) Dose-response relationship from longitudinal data with response-dependent dose-modification using likelihood methods. Biometrical J 54:494–506

Funatogawa I, Funatogawa T, Ohashi Y (2007) An autoregressive linear mixed effects model for the analysis of longitudinal data which show profiles approaching asymptotes. Stat Med 26:2113–2130

Funatogawa I, Funatogawa T, Ohashi Y (2008a) A bivariate autoregressive linear mixed effects model for the analysis of longitudinal data. Stat Med 27:6367–6378

Funatogawa T, Funatogawa I, Takeuchi M (2008b) An autoregressive linear mixed effects model for the analysis of longitudinal data which include dropouts and show profiles approaching asymptotes. Stat Med 27:6351–6366

Lindsey JK (1993) Models for repeated measurements. Oxford University Press

Rabe-Hesketh S, Skrondal A (2012) Multilevel and longitudinal modeling using Stata. Continuous responses, vol I, 3rd edn. Stata Press

Rosner B, Muñoz A (1988) Autoregressive model for the analysis of longitudinal data with unequally spaced examinations. Stat Med 7:59–71

Rosner B, Muñoz A (1992) Conditional linear models for longitudinal data. In: Dwyer JM, Feinleib M, Lippert P, Hoffmeister H (eds) Statistical models for longitudinal studies of health. Oxford University Press, pp 115–131

Rosner B, Muñoz A, Tager I, Speizer F, Weiss S (1985) The use of an autoregressive model for the analysis of longitudinal data in epidemiology. Stat Med 4:457–467

Schmid CH (1996) An EM algorithm fitting first-order conditional autoregressive models to longitudinal data. J Am Stat Assoc 91:1322–1330

Schmid CH (2001) Marginal and dynamic regression models for longitudinal data. Stat Med 20:3295–3311

Chapter 3
Case Studies of Autoregressive Linear Mixed Effects Models: Missing Data and Time-Dependent Covariates

Abstract In the previous chapter, we introduced autoregressive linear mixed effects models for analysis of longitudinal data. In this chapter, we provide examples of actual data analysis using these models. We also discuss two topics from the medical field: response-dependent dropouts and response-dependent dose modifications. When the missing mechanism depends on the observed, but not on the unobserved, responses, it is termed missing at random (MAR). The missing process does not need to be simultaneously modeled for the likelihood because the likelihood can be factorized into two parts: one for the measurement process and the other for the missing process. Maximum likelihood estimators are consistent under MAR if the joint distribution of the response vector is correctly specified. For the problem of dose modification, similar concepts are applied. When the dose modification depends on the observed, but not on the unobserved, responses, the dose process does not need to be simultaneously modeled for the likelihood. Here, we analyze schizophrenia data and multiple sclerosis data using autoregressive linear mixed effects models as examples of response-dependent dropouts and response-dependent dose modifications, respectively.

Keywords Autoregressive linear mixed effects model · Dose modification Longitudinal · Missing · Time-dependent covariate

3.1 Example with Time-Independent Covariate: PANSS Data

In this and next sections, we analyze data obtained from a schizophrenia trial (Marder and Meibach 1994). This is a randomized controlled trial composed of placebo, haloperidol, and risperidone groups. Although the risperidone group has four dose levels, we combine the four dose levels into one as Diggle et al. (2002) did. The primary response variable was the total score obtained on the positive and negative symptom rating scale (PANSS). A higher score indicates a worse condition, and the

© The Author(s), under exclusive licence to Springer Nature Singapore Pte Ltd. 2018
I. Funatogawa and T. Funatogawa, *Longitudinal Data Analysis*, JSS Research Series in Statistics, https://doi.org/10.1007/978-981-10-0077-5_3

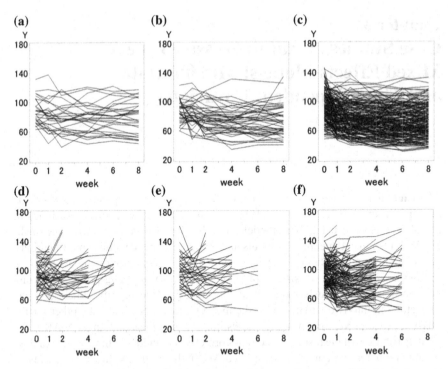

Fig. 3.1 Observed PANSS longitudinal data for each subject. **a–c** Completers, **d–f** subjects dropped out. **a, d** Placebo, **b, e** haloperidol, **c, f** risperidone

drugs are expected to decrease the score. Treatment groups are a time-independent covariate. The data are illustrated by Diggle et al. (2002).

The numbers of subjects who were measured just before and after the adminis-trations were 85 for the placebo group, 85 for the haloperidol group, and 336 for the risperidone group. Figure 3.1 shows the profiles of each subject in each group. The data are divided into completers and subjects dropped out. In dropouts, all data after some time points for each subject are missing. Although the PANSS score was measured at weeks −1, 0, 1, 2, 4, 6, and 8 after initial administration, we do not use the scores at week −1 in our analysis. The dropout proportions were 65.9% for the placebo group, 51.8% for the haloperidol group, and 42.0% for the risperidone group. The proportions differ among the three groups with the placebo group having the highest dropout proportion. The most common reason for dropouts was inadequate response (Diggle et al. 2002).

The schizophrenia trial data were analyzed as an example of data including dropouts (Diggle et al. 2002; Funatogawa et al. 2008b; Henderson et al. 2000; Xu and Zeger 2001). Quadratic polynomial time trend models were sometimes assumed for the PANSS scores (Diggle et al. 2002; Xu and Zeger 2001). Diggle et al. (2002) indicated that nonlinear models that express a profile approaching an asymptote may be preferable on biological grounds.

Funatogawa et al. (2008b) adopt the following autoregressive linear mixed effects model to the schizophrenia trial data,

$$\begin{cases} Y_{i,0} = \beta_{g,\text{base}} + b_{\text{base}\,i} + \varepsilon_{(\text{AR,ST})i,0} + \varepsilon_{(\text{ME})i,0} \\ Y_{i,t} = \rho Y_{i,t-1} + \beta_{g,\text{int}} + b_{\text{int}\,i} + \varepsilon_{(\text{AR,ST})i,t} + \varepsilon_{(\text{ME})i,t} - \rho\varepsilon_{(\text{ME})i,t-1}, \; (t > 0) \end{cases},$$

$$(3.1.1)$$

where g is a subscript representing the three treatment groups (1, placebo; 2, haloperidol; 3, risperidone). The random baseline, $b_{\text{base}\,i}$, and the random follow-up intercept, $b_{\text{int}\,i}$, are assumed to be normally distributed with the mean zero, variances, σ_{G0}^2 and σ_{G1}^2, and covariance, σ_{G01}. $\varepsilon_{(\text{AR,ST})i,0}$, $\varepsilon_{(\text{AR,ST})i,t}$, and $\varepsilon_{(\text{ME})i,t}$ are independently normally distributed with the mean zero and variances $\left(1 - \rho^2\right)^{-1}\sigma_{\text{AR,ST}}^2$, $\sigma_{\text{AR,ST}}^2$, and σ_{ME}^2. Here, AR, ST means stationary AR(1) and ME means a measurement error. The responses are expected to approach asymptotes $\left(\beta_{g,\text{int}} + b_{\text{int}\,i}\right)/(1 - \rho)$. $b_{\text{int}\,i}/(1 - \rho)$ is the random effect in the asymptotes. The model is the three-group version of the model (2.2.6) with the error structure (2.4.3).

We analyze the data with $t=0$, 1, 2, 3, 4, and 5. We use the likelihood of the autoregressive form in Sect. 2.5.2 and the SAS code of MIXED procedure for indirect methods in Sect. 2.5.3.

The estimates of the fixed parameters are $\beta_{1,\text{base}} = 92.4$, $\beta_{2,\text{base}} = 93.6$, $\beta_{3,\text{base}} = 92.4$, $\beta_{1,\text{int}} = 50.4$, $\beta_{2,\text{int}} = 46.7$, $\beta_{3,\text{int}} = 42.7$, and $\rho = 0.451$. The asymptotes, $\beta_{g,\text{int}}/(1 - \rho)$, are 91.8, 85.1, and 77.8, respectively. The estimates for the covariance are $\sigma_{G0}^2 = 199.0$, $\sigma_{G1}^2 = 133.5$, $\sigma_{G01} = 107.7$, $\sigma_{\text{AR,ST}}^2 = 127.0$, and $\sigma_{\text{ME}}^2 = 0.4066$. The variance of the asymptote is $\sigma_{G1}^2/(1 - \rho)^2 = 442.9$. If σ_{ME}^2 is ignored in $\mathbf{R}_i$ because σ_{ME}^2 is much smaller than σ_{AR}^2, the change in -2 log-likelihood is less than 0.1. Therefore, $\mathbf{R}_i$ would be expressed only by an AR(1) error.

We assume a stationary AR(1) error because it is popular in longitudinal data analysis. However, it may be more natural to assume that the element of the AR(1) error at baseline is zero. In this case, the marginal variance covariance matrix of the response and estimates of the fixed effects are not changed. The variance of the random baseline effect changes from 199.0 to 358.5 and the correlation between the random baseline and the random asymptote changes from 0.66 to 0.48. The assumption affects the Bayes predictions.

3.2 Missing Data

3.2.1 Missing Mechanism

In the schizophrenia trial data, many subjects dropped out. In the medical research with human subject, problems of missing or dropouts in longitudinal data occur frequently. The problem of missing responses in longitudinal data analyses has been studied extensively in the past several decades. In the problem, it is discussed whether the missing process needs to be simultaneously modeled with the measurement process.

Let $\mathbf{r}$ be a vector of indicator variables for the missing process, such that $r_{i,t} = 1$ if $Y_{i,t}$ is observed, and $r_{i,t} = 0$ otherwise. Given $\mathbf{r}$, the complete response vector $\mathbf{Y}$ can be partitioned into $\mathbf{Y} = (\mathbf{Y}_{\text{obs}}, \mathbf{Y}_{\text{miss}})$, where $\mathbf{Y}_{\text{obs}}$ is the observed responses and $\mathbf{Y}_{\text{miss}}$ is the unobserved responses. We express $\mathbf{Y}$ and $\mathbf{r}$ as

$$f(\mathbf{Y}, \mathbf{r}|\mathbf{x}, \boldsymbol{\theta}, \boldsymbol{\xi}) = f(\mathbf{Y}|\mathbf{x}, \boldsymbol{\theta}) f(\mathbf{r}|\mathbf{Y}, \mathbf{x}, \boldsymbol{\xi}), \qquad (3.2.1)$$

where $\boldsymbol{\theta}$ and $\boldsymbol{\xi}$ denote parameters for the measurement process and the missing process, respectively, and $\mathbf{x}$ denotes covariates.

Based on the missing process, $f(\mathbf{r}|\mathbf{Y}, \mathbf{x}, \boldsymbol{\xi}) = f(\mathbf{r}|\mathbf{Y}_{\text{obs}}, \mathbf{Y}_{\text{miss}}, \mathbf{x}, \boldsymbol{\xi})$, we can hierarchically classify missing mechanisms into the following three levels: missing completely at random (MCAR), missing at random (MAR), and missing not at random (MNAR) (Little and Rubin 1987; Laird 1988).

When the missing mechanism does not depend on the measurement process as

$$f(\mathbf{r}|\mathbf{Y}_{\text{obs}}, \mathbf{Y}_{\text{miss}}, \mathbf{x}, \boldsymbol{\xi}) = f(\mathbf{r}|\mathbf{x}, \boldsymbol{\xi}), \qquad (3.2.2)$$

it is termed MCAR. Usual standard analytical approaches provide consistent estimators under MCAR.

When the missing mechanism depends on the observed responses, but not on the unobserved responses,

$$f(\mathbf{r}|\mathbf{Y}_{\text{obs}}, \mathbf{Y}_{\text{miss}}, \mathbf{x}, \boldsymbol{\xi}) = f(\mathbf{r}|\mathbf{Y}_{\text{obs}}, \mathbf{x}, \boldsymbol{\xi}), \qquad (3.2.3)$$

it is termed MAR. The missing process does not need to be simultaneously modeled for the likelihood because the likelihood can be factorized into two parts for the measurement process and the missingness process. The maximum likelihood estimators (MLEs) are consistent under MAR if the joint distribution of the response vector is correctly specified. However, non-likelihood-based methods, which do not specify the joint distribution of the response vector such as a generalized estimating equation (GEE) method, provide biased estimators under MAR.

When the missing mechanism depends on unobserved responses, that is MNAR, MLEs that ignore missing mechanisms are biased. The missing process needs to be simultaneously modeled with the measurement process, and several approaches have been proposed. Selection models are based on the factorization (3.2.1) of the missing process given the measurements $\mathbf{Y}$ and the measurement process. In contrast, the pattern mixture models are based on the factorization of the measurement process given the missing pattern and the missing process. In the shared parameter models, random effects are shared in both the measurement process and missing process.

3.2.2 Model Comparison: PANSS Data

In the schizophrenia trial data presented in Sect. 3.1, the scores of some patients increased just before dropouts as shown in Fig. 3.1. This feature resembles the results of the simulation study under MAR dropouts (Funatogawa et al. 2008b). Some subjects seem to dropout directly based on their observed values. The assumption of MAR may be reasonable. In this section, we compare several models with different mean and variance covariance structures to assess the influence of missing data.

For balanced data with no missing data, estimates of the discrete means at each time point in each group are the same, irrespective of the variance covariance structure, and the differences in the variance covariance structures affect the standard errors. If there are missing data, the estimates of means differ depending on the variance covariance structure.

We compare the autoregressive linear mixed effects model (3.1.1) and the model without the measurement error with several marginal models and linear mixed effects models introduced in Chap. 1, and summarize them in Table 3.1. Table 3.1 shows the number of parameters for the mean structure and variance covariance structure, along with Akaike's information criterion (AIC) for each model. We examine two assumptions for each model, one in which the variance covariance matrices $\mathbf{V}_i$ are the same across all three groups and one in which they differ across the three groups. In the autoregressive linear mixed effects model (3.1.1), the number of parameters for the mean structure and variance covariance structure are 7 and 6, respectively. Because the autoregressive parameter ρ is used in both mean and variance covariance structures, the total number of parameters is 12. Under the different variance covariance matrices, ρ differs across the three groups and the number of parameters in the mean structure increases by 2.

We examine the marginal models with the discrete means at each time point in each group,

$$\begin{cases} Y_{ij} = \mu_{gj} + \varepsilon_{ij} \\ \boldsymbol{\varepsilon}_i \sim \mathrm{MVN}(\mathbf{0}, \mathbf{R}_i) \end{cases}, \tag{3.2.4}$$

where $\boldsymbol{\varepsilon}_i = \left(\varepsilon_{i1}, \varepsilon_{i2}, \cdots, \varepsilon_{in_i}\right)^T$. This model is the three-group version of the model (1.3.9) with some error structures. We examine several variance covariance structures: unstructured (UN), independent equal variance, independent unequal variance, AR(1), heterogenous AR(1) (ARH(1)), compound symmetry (CS), heterogenous CS (CSH), Toeplitz, heterogenous Toeplitz, first-order ante-dependence (ANTE(1)), and ANTE(1) and a random intercept. The details of these structures are given in Sect. 1.4.2 and Table 1.1. The discrete means with the CS correspond to means at each time point with a random intercept, as discussed in Sects. 1.3.1 and 1.3.2. The discrete means with the UN is discussed in Sect. 1.3.3. The discrete means with the UN may be used as a reference because there are no constrains on the mean and variance covariance parameters. However, the UN is not parsimonious using $6(6+1)/2 = 21$ parameters under the same variance covariance matrices across the groups and $21 \times 3 = 63$ parameters under the different variance covariance matrices.

Table 3.1 Model comparison for PANSS data

Model	Same V_i across groups			Different V_i across groups		
Mean structures Variance covariance structures	NOP[a] for mean	NOP for variance covariance	AIC	NOP[a] for mean	NOP for variance covariance	AIC
(a) Autoregressive linear mixed effects model of baseline and asymptote						
Random baseline and asymptote and $\sigma^2_{AR,ST}$ and σ^2_{ME}	7	5(6)[b]	19894.3	9	15(18)[b]	19898.8
Random baseline and asymptote and $\sigma^2_{AR,ST}$	7	4(5)	19892.3	9	12(15)	19894.3
(b) Means at each time in each group[c]						
Unstructured (UN)	18	21	19895.5	18	63	19902.3
Independent equal variance	18	1	21714.4	18	3	21717.0
Independent unequal variance	18	6	21709.1	18	18	21726.2
AR(1)	18	2	19994.2	18	6	20001.8
Heterogeneous AR(1)	18	7	20000.6	18	21	20008.8
Compound symmetry (CS)	18	2	20318.3	18	6	20324.5
Heterogeneous CS	18	7	20311.4	18	21	20322.1
Toeplitz	18	6	19985.6	18	18	19999.3
Heterogeneous Toeplitz	18	11	19992.4	18	33	20008.4
First-order ante-dependence (ANTE(1))	18	11	19901.5	18	33	19897.6
ANTE(1) and random intercept	18	12	19896.0	18	36	[d]
(c) Linear mixed effects model of linear time trends						
Random intercept	6	2	20410.7	6	6	20415.5
Random intercept and slope	6	4	20245.0	6	12	20256.7
(d) Linear mixed effects model of quadratic time trends						
Random intercept, time, time^2	9	7	20036.6	9	21	20052.1

[a] NOP: number of parameters. [b] The number in parentheses include the autoregressive parameters that also appear in the mean structure. [c] See variance covariance structures given in Sect. 1.4.2 and Table 1.1. [d] Not calculated

We examine linear mixed effects models of linear time trends with only a random intercept and with a random intercept and slope. The latter model is the three-group version of the model (1.3.12). We also examine linear mixed effects models of quadratic time trends with a random intercept, time, and time^2,

$$
\begin{cases}
Y_{ij} = (\beta_{g0} + b_{0i}) + (\beta_{g1} + b_{1i})t_{ij} + (\beta_{g2} + b_{2i})t_{ij}^2 + \varepsilon_{ij} \\[2mm]
\begin{pmatrix} b_{0i} \\ b_{1i} \\ b_{2i} \end{pmatrix} \sim \text{MVN}\left(\begin{pmatrix} 0 \\ 0 \\ 0 \end{pmatrix}, \begin{pmatrix} \sigma_{G0}^2 & \sigma_{G01} & \sigma_{G02} \\ \sigma_{G01} & \sigma_{G1}^2 & \sigma_{G12} \\ \sigma_{G02} & \sigma_{G12} & \sigma_{G2}^2 \end{pmatrix} \right) \\[2mm]
\varepsilon_{ij} \sim \text{N}(0, \sigma_\varepsilon^2)
\end{cases}
\tag{3.2.5}
$$

The models with the same variance covariance matrices across groups were better than the models with the different variance covariance matrices in all models based on the AICs, so we discuss the results of the former. The autoregressive linear mixed effects model with a random baseline and asymptote and AR(1) error showed the best fit with AIC = 19,892.3. Next to the two autoregressive linear mixed effects models, the discrete means with the UN showed good fit with AIC = 19,895.5. The discrete means with the ANTE(1) and a random intercept (AIC = 19,896.0) was slightly worse than the discrete means with the UN but showed a comparable fit. The discrete means with the AR(1), CS, or Toeplitz or the heterogeneous variance versions of these structures did not show good fit. Linear mixed effects models of linear or quadratic time trends did not show good fits.

Table 3.2 shows the estimates of marginal variances, covariances, and correlations for the autoregressive linear mixed effects model with a random baseline and asymptote and AR(1) error and the discrete means with the UN (Funatogawa et al. 2008b). The estimate of the marginal variance covariance matrix of the autoregressive linear mixed effects model is given by $\Sigma_i = (I_i - \rho F_i)^{-1}(Z_i G Z_i^T + R_i)\{(I_i - \rho F_i)^{-1}\}^T$ (2.3.17). The estimates of the UN with 21 parameters were similar to those of the autoregressive linear mixed effects model with 5 parameters. The variances increased with time but attenuated at the end. Let corr$_{j,k}$ be the j,kth correlation. Considering the correlations, corr$_{j,j-l}$ ($j = l+1, l+2, \cdots, 6$), with the fixed time intervals l, viewing diagonally, these were not constant. The correlation was larger for the later time j, and the correlation with the first time point, corr$_{1+l,1}$, was particularly lower than the other correlations corr$_{j,j-l}$. Considering the correlations, corr$_{j,j-l}$ ($l = 1, 2, \cdots, j-1$), with the fixed time j, viewing vertically, the correlation was smaller for the longer time interval l.

Figure 3.2 shows estimated marginal variance covariance matrices for (a) the autoregressive linear mixed effects model with a random baseline and asymptote and AR(1) error, (b) UN, (c) CS, (d) heterogeneous Toeplitz, (e) ANTE(1) and a random intercept, and (f) linear mixed effects model of quadratic time trends with a random intercept, time, and time^2. The darker color indicates the higher variance or covariance value. All covariances were positive. The matrices (a), (b), (e), and (f) can express non-stationary structures. These showed similar matrices and expressed the data well. However, the numbers of parameters of the UN and the ANTE(1)

Table 3.2 Estimates of marginal variance, covariance, and correlation for PANSS data

Autoregressive linear mixed effects model with random baseline and asymptote and $\sigma^2_{AR,ST}$[a]	Discrete means with unstructured (UN)[a]
360 270 230 210 200 200	360 250 240 230 220 190
.69 430 380 350 340 340	.64 430 380 360 340 320
.53 .80 510 450 420 400	.56 .82 500 450 430 390
.47 .72 .83 560 480 450	.51 .73 .84 570 510 480
.44 .68 .76 .85 580 500	.46 .65 .76 .86 620 560
.43 .66 .73 .78 .85 590	.41 .61 .70 .81 .89 620

Funatogawa et al. (2008b)

[a]Covariances are above the diagonal, variances are on the diagonal, and correlations are below the diagonal

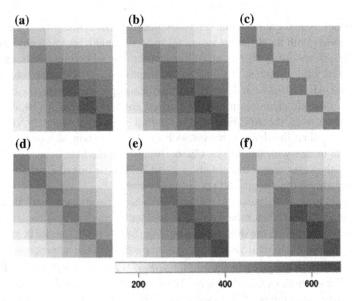

Fig. 3.2 Variance covariance matrices of PANSS. The darker color shows the higher value. **a** Autoregressive linear mixed effects model with a random baseline and asymptote and AR(1) error, **b** discrete means with unstructured, **c** discrete means with compound symmetry, **d** discrete means with heterogeneous Toeplitz, **e** discrete means with ANTE(1) and a random intercept, **f** linear mixed effects model of quadratic time trends with a random intercept, time, and time^2

and a random intercept increase with the number of time points and these are not parsimonious. The CS and heterogeneous Toeplitz showed different values from these matrices. These stationary correlation structures in which the correlations with fixed time intervals are the same were unable to express the data well. The CS whose variances were the same across time points was too simple.

Figure 3.3 shows how the assumed models affect the estimates of means at each time point in each group. The models examined are (a) the autoregressive linear mixed effects model with a random baseline and asymptote and AR(1) error,

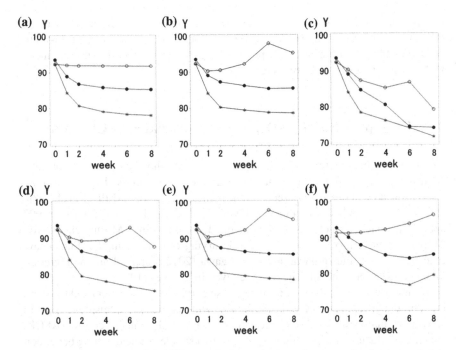

Fig. 3.3 Estimated means of PANSS at each time point in each group: placebo (open circle), haloperidol (closed circle), and risperidone (asterisk). **a** Autoregressive linear mixed effects model with a random baseline and asymptote and AR(1) error, **b** discrete means with unstructured, **c** discrete means with independent structures (simple means), **d** discrete means with compound symmetry, **e** discrete means with ANTE(1) and a random intercept, **f** linear mixed effects model of quadratic time trends with a random intercept, time, and time^2

(b) discrete means with the UN, (c) discrete means with independent structures, (d) discrete means with the CS, (e) discrete means with the ANTE(1) and a random intercept, and (f) linear mixed effects model of quadratic time trends with a random intercept, time, and time^2. The open circles, the closed circles, and the asterisk show the placebo group, the haloperidol group, and the risperidone group, respectively.

The MLEs are consistent under MAR if the joint distribution of the response vector is correctly specified. The discrete means with independent structures, irrespective of equal or unequal variance assumptions, provide the observed mean response profiles which are the simple means of only observed values at each time point in each group. The responses from a subject have usually positive correlations in longitudinal data analysis, and the independent assumption is obviously incorrect. Because the subjects with higher scores dropped out in this example, the estimates were biased downward. On the other hand, the discrete means with the UN has the least assumption and the bias is expected to be small under MAR. Compared with the discrete means with the UN, the discrete means with independent structures were lower in each group as expected. Although not as much as independent structures, the discrete means with the CS also showed lower means compared with the means with the UN. The autoregressive linear mixed effects models and the discrete means with the ANTE(1) and a random intercept provided higher values than the discrete means with independent structures or the CS.

3.3   Example with Time-Dependent Covariate: AFCR Data

As an example of a time-dependent covariate, we show data from a placebo-controlled, randomized, double-masked, variable dosage, clinical trial of azathioprine with and without methylprednisolone in multiple sclerosis (Ellison et al. 1989). Heitjan (1991) applied a nonlinear growth curve to the data, and Lindsey (1993) applied a linear mixed effects model of a quadratic time trend. Misumi and Konishi (2016) applied a mixed effects historical varying-coefficient model. Funatogawa et al. (2007) and Funatogawa and Funatogawa (2012a) applied an autoregressive linear mixed effects model. The data are given in Lindsey (1993).

We introduce the analysis of dose-response curves for each patient and the population in the group that received azathioprine without methylprednisolone (Funatogawa and Funatogawa 2012a). The response variable is absolute F_c receptor (AFCR), which is a measure of the immune system with a smaller value denoting better conditions. The dose at the start, 2.2 mg/kg daily, is defined as one unit. Doses were modified throughout the trial. Treatment continued for up to 4 years. AFCR was measured prior to initiation of therapy, at initiation, at weeks 4, 8, and 12, and every 12 weeks thereafter. We followed Heitjan (1991) in using a square root transformation on the AFCR responses. The number of patients was 15, and the total number

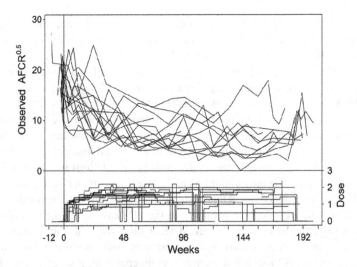

Fig. 3.4 Observed $AFCR^{0.5}$ and dose in each patient. Funatogawa and Funatogawa (2012a)

of the measurements was 244. Figure 3.4 shows the $\text{AFCR}^{0.5}$ and dose profiles for each patient. The timings of dose modification and measurement were irregular and differ among patients, as shown in the figure.

We applied the autoregressive linear mixed effects model (2.2.8) in Sect. 2.2.3 with the error structure (2.4.1) to the multiple sclerosis data,

$$
\begin{cases}
Y_{i,0} = \beta_{\text{base}} + b_{\text{base}\,i} + \varepsilon_{i,0} \\
Y_{i,t} = \rho Y_{i,t-1} + \beta_{\text{int}} + b_{\text{int}\,i} + (\beta_{\text{cov}} + b_{\text{cov}\,i})x_{i,t} + \varepsilon_{i,t}, (t > 0) \\
\varepsilon_{i,0} = \varepsilon_{(\text{ME})i,0} \\
\varepsilon_{i,t} = \varepsilon_{(\text{AR})i,t} + \varepsilon_{(\text{ME})i,t} - \rho\varepsilon_{(\text{ME})i,t-1}, (t > 0)
\end{cases}
\tag{3.3.1}
$$

where $Y_{i,t}$ and $x_{i,t}$ are the $\text{AFCR}^{0.5}$ level and the dose of the ith subject at time t, respectively. $b_{\text{cov}\,i}$, a coefficient of the covariate $x_{i,t}$, is a random variable. The asymptote of the subject i at time t, $Y_{\text{Asy}\,i,t}$, is

$$
Y_{\text{Asy}\,i,t} = (1 - \rho)^{-1}\{\beta_{\text{int}} + b_{\text{int}\,i} + (\beta_{\text{cov}} + b_{\text{cov}\,i})x_{i,t}\}.
\tag{3.3.2}
$$

The asymptote depends on the covariate $x_{i,t}$. The term $(1 - \rho)^{-1}b_{\text{cov}\,i}$ represents the difference in sensitivity to dose modifications across subjects. Both the fixed and random effects have the baseline, follow-up intercept, and follow-up dose effect. The random effects $\mathbf{b}_i = (b_{\text{base}\,i}, b_{\text{int}\,i}, b_{\text{cov}\,i})^T$ are assumed to be normally distributed with the mean zero and unstructured variance covariance matrix $\mathbf{G}$. $\varepsilon_{(\text{AR})i,t}$ and $\varepsilon_{(\text{ME})i,t}$ are independently normally distributed with the mean zero and variances σ_{AR}^2 and σ_{ME}^2.

We used the state space representation and Kalman filter in Chap. 6 to obtain the likelihood. SAS IML procedure is used for the calculation and optimization.

Table 3.3 Parameter estimates, covariance, standard deviation, and correlation for the multiple sclerosis data

Parameter	Estimate	SE	Covariance, standard deviation and correlation for random effects[a]		
				$b_{\text{base}\,i}$	$\dfrac{b_{\text{int}\,i}}{1-\rho}$	$\dfrac{b_{\text{cov}\,i}}{1-\rho}$
β_{base}	17.0	0.9			
$(1-\rho)^{-1}\beta_{\text{int}}$	9.6	1.3	$b_{\text{base}\,i}$	3.15	−0.62	4.71
$(1-\rho)^{-1}\beta_{\text{cov}}$	−2.2	1.2				
ρ^{84}	0.63	0.05	$\dfrac{b_{\text{int}\,i}}{1-\rho}$	−0.07	2.65	−7.59
σ_{ME}	2.48					
$\sigma_{\text{AR}}(1-\rho^2)^{-0.5}$	1.70		$\dfrac{b_{\text{cov}\,i}}{1-\rho}$	0.48	−0.91	3.13

Funatogawa and Funatogawa (2012a)

[a]Covariance are above the diagonal, standard deviations are on the diagonal, and correlations are below the diagonal

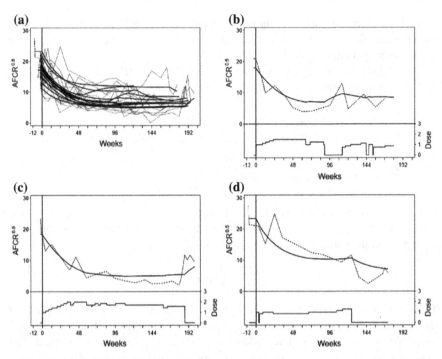

Fig. 3.5 Fitted and observed AFCR$^{0.5}$. **a** Solid and dashed lines are fitted and observed AFCR$^{0.5}$, respectively, in each patient. **b–d** Three representative patients. Upper and lower solid lines indicate fitted AFCR$^{0.5}$ and dose profiles, respectively. A dashed line indicates observed AFCR$^{0.5}$. Funatogawa and Funatogawa (2012a)

Table 3.3 shows the parameter estimates. Figure 3.5a shows the fitted and observed values of AFCR$^{0.5}$ in all patients, and Fig. 3.5b–d shows fitted and observed values with actual doses in three representative patients. The fitted values were calculated from the predicted values of the random effects and actual doses for each patient. The model well represents the gradual decreasing response levels and inter-individual differences at baseline and later time points. The estimate of the autoregressive coefficient for a time unit of 12 weeks is $\hat{\rho}^{84} = 0.627$. Patients usually showed an increasing response level after stopping drug administration. Two patients, including the patient in Fig. 3.5d, however, showed a decreasing response level after stopping drug administration. Figure 3.6 shows the dose-response curves of the asymptotes in each patient and the population mean with the estimates of baseline. The estimates of population means are 17.0 for the baseline, and 9.6, 7.4, and 5.2 for the asymptotes at doses of 0, 1, and 2 units, respectively. This figure represents well the dose-response curves of the population mean and its inter-individual difference. The asymptotes decrease according to the dose in all patients except two patients mentioned above.

Based on the Akaike information criteria (AIC), each of $\varepsilon_{(AR)i,t}$, $\varepsilon_{(ME)i,t}$, and random effects improves the fit, and there exists an inter-individual variability of

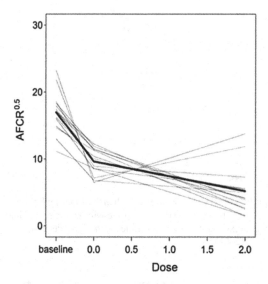

Fig. 3.6 Estimated dose-response curves of $AFCR^{0.5}$ asymptotes with estimated baseline. Thin and thick lines indicate the estimates for each patient and the population mean, respectively. Funatogawa and Funatogawa (2012a)

the dose effect and a significant dose effect as a population average. When random effects are replaced with a random intercept, the fit is worse. If the previous response is excluded from the covariates, that is $\rho = 0$, and $\varepsilon_{(AR)i,t}$ is replaced with a stationary AR(1) error, the fit is obviously worse. This is a linear mixed effects model. It assumes that the current response depends on the current dose but not on the previous doses, and the response changes to a new level without delay. In contrary, the response changes to a new level gradually in the model (3.3.1).

We briefly introduce the analysis of three group comparison (Funatogawa et al. 2007). The three groups are azathioprine with methylprednisolone, azathioprine without methylprednisolone, and placebo. The autoregressive linear mixed effects model (3.3.1) with $\beta_{g,\text{base}}$, $\beta_{g,\text{int}}$, and $\beta_{g,\text{cov}}$ instead of β_{base}, β_{int}, and β_{cov} is applied. For comparison, the following linear mixed effects model with a quadratic time trend is also applied,

$$Y_{ij} = \beta_{g0} + \beta_{g1}t_{ij} + \beta_{g2}t_{ij}^2 + \beta_{g3}x_{ij} + \beta_{g4}x_{ij}t_{ij} + b_{0i} + \varepsilon_{e(AR)ij} + \varepsilon_{(ME)ij}.$$

Figure 3.7 shows the expected profiles of AFCR for each group based on each model when the doses are 1.5 throughout the trial. The parameters in the quadratic time trend are hard to interpret.

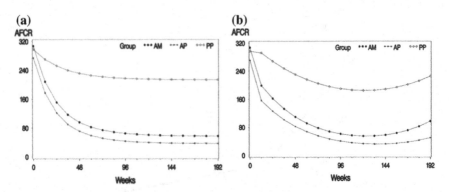

Fig. 3.7 Expected profiles of AFCR for each group when the doses are 1.5 throughout the trial. AM, azathioprine with methylprednisolone; AP, azathioprine without methylprednisolone; PP, placebo. **a** Autoregressive linear mixed effects model. **b** Linear mixed effects model with a quadratic time trend. Funatogawa et al. (2007)

3.4 Response-Dependent Modification of Time-Dependent Covariate

The previous section provides an example of autoregressive linear mixed effects models with a time-dependent covariate. Section 4.3 provides an example of analysis based on bivariate autoregressive linear mixed effects models with a time-dependent covariate (Funatogawa et al. 2008a), which is an example of response-dependent dose modification. Active vitamin D_3 is administered repeatedly for the treatment of secondary hyperparathyroidism, and it decreases parathyroid hormone (PTH) levels. The dose is adjusted in each patient according to the PTH and Ca levels and other medical conditions. The PTH and Ca levels have target ranges defined by a clinical guideline (KDOQI 2007). Even under these situations, we can estimate the dose-response relationship.

For dose modification, Funatogawa and Funatogawa (2012b) applied similar concepts with missing data mechanism in Sect. 3.2. Here, $Y_{i,0}$ is a baseline measurement, and $Y_{i,t}$ is the tth $(t = 1, \cdots, T_i)$ measurement after the baseline measurement in the ith $(i = 1, \cdots, N)$ subject. $X_{i,t}$ is the dose between $t - 1$ and t. $Y_{i,t}$ is measured after the administration of $X_{i,t}$. For simplicity, we consider no other time-dependent covariates. $\mathbf{Z}_i$ is the vector of time-independent covariates.

The joint density function of $\mathbf{Y}_{i,T_i}^{(h)} = (Y_{i,0}, Y_{i,1}, \cdots, Y_{i,T_i})^T$ and $\mathbf{X}_{i,T_i}^{(d)} = (X_{i,1}, X_{i,2}, \cdots, X_{i,T_i})^T$ is expressed as $f\left(\mathbf{Y}_{i,T_i}^{(h)}, \mathbf{X}_{i,T_i}^{(d)} | \mathbf{Z}_i, \boldsymbol{\theta}, \boldsymbol{\Psi}\right)$, where $\boldsymbol{\theta}$ and $\boldsymbol{\Psi}$ denote parameters for the measurement process and the dose process, respectively. The superscripts (h) and (d) indicate the history of the responses and doses. We are interested in the inferences regarding $\boldsymbol{\theta}$, particularly the parameters that show how the response changes with the dose. Taking into account the sequence between $Y_{i,t}$ and $X_{i,t}$, we factorize the likelihood corresponding to the ith subject,

$L_i(\theta, \Psi) = f\left(\mathbf{Y}_{i,T_i}^{(h)}, \mathbf{X}_{i,T_i}^{(d)} | \mathbf{Z}_i, \theta, \Psi\right)$, into two parts: the likelihood for the measurement process,

$$L_i(\theta) = f\left(Y_{i,0} | \mathbf{Z}_i, \theta\right) \prod_{t=1}^{T_i} f\left(Y_{i,t} | \mathbf{Y}_{i,t-1}^{(h)}, \mathbf{X}_{i,t}^{(d)}, \mathbf{Z}_i, \theta\right), \qquad (3.4.1)$$

and the likelihood for the dose process,

$$L_i(\Psi) = f\left(X_{i,1} | Y_{i,0}, \mathbf{Z}_i, \Psi\right) \prod_{t=2}^{T_i} f\left(X_{i,t} | \mathbf{Y}_{i,t-1}^{(h)}, \mathbf{X}_{i,t-1}^{(d)}, \mathbf{Z}_i, \Psi\right). \qquad (3.4.2)$$

The likelihood is $L_i(\theta, \Psi) = L_i(\theta) \times L_i(\Psi)$.

If the parameters θ and Ψ are separable in the above factorization, maximum likelihood methods based on $L_i(\theta)$ alone are valid for the inferences regarding θ without modeling the dose process. For example, θ and Ψ are separable if the next dose is selected based on observed responses and the new response is generated by the administered doses. This is similar to MAR. If the next dose depends on unobserved parts of the measurement process, such as unknown parameters, given the observed responses or the new response is generated by unobserved parts of the dose process, θ and Ψ are not separable. In this case, the two processes have several common parameters, and the dose process needs to be simultaneously modeled with the measurement process. This is similar to MNAR.

Diggle et al. (2002) showed similar factorization of likelihood for transition models of binary responses. However, Eq. (3.4.1) can be applied to more general models, such as mixed effects models (Laird and Ware 1982; Funatogawa et al. 2007). Maximum likelihood methods for standard models can provide estimates of the dose-response without modeling the dose process if the dose modifications are based on observed responses and the assumed model of the measurement process is correct (Funatogawa and Funatogawa, 2012b). As a merit of mixed effects models, these provide estimates of each patient's dose-response curve. The dose modification based on observed responses may be prespecified in a study protocol.

In the area of causal modeling in epidemiology, as another approach, marginal structural models with an inverse probability of treatment weighted (IPTW) estimators were used for flexible dose studies (Lipkovich et al. 2008). These require to model the treatment process and to assume the probability of receiving each treatment is bounded away from zero. If the treatment modification is determined according to the same criterion, this assumption does not hold. Based on simulation studies for flexible dose titration, the autoregressive linear mixed effects models may be an appropriate modeling option in identifying the dose-response compared with the marginal models (Xu et al. 2012).

Another example of response-dependent dose modification is erythropoietin for the treatment of renal anemia in hemodialysis patients. The drug is repeatedly administered and raises the hemoglobin (Hb) levels. A clinical guideline defines a target range of Hb levels (National Kidney Foundation 2003). The doses are adjusted according to Hb levels. As an example of a time-dependent covariate beside the dose, the relationship between blood donation and Hb levels are analyzed by applying autoregressive linear mixed effects models (Nasserinejad et al. 2016).

References

Diggle PJ, Heagerty P, Liang K-Y, Zeger SL (2002) Analysis of longitudinal data (2nd edn). Oxford University Press

Ellison GW, Myers LW, Mickey MR, Graves MC, Tourtellotte WW, Syndulko K, Holevoet-Howson MI, Lerner CD, Frane MV, Pettler-Jennings P (1989) A placebo-controlled, randomized, double-masked, variable dosage, clinical trial of azathioprine with and without methylprednisolone in multiple sclerosis. Neurology 39:1018–1026

Funatogawa I, Funatogawa T (2012a) An autoregressive linear mixed effects model for the analysis of unequally spaced longitudinal data with dose-modification. Stat Med 31:589–599

Funatogawa I, Funatogawa T (2012b) Dose-response relationship from longitudinal data with response-dependent dose-modification using likelihood methods. Biometrical J 54:494–506

Funatogawa I, Funatogawa T, Ohashi Y (2007) An autoregressive linear mixed effects model for the analysis of longitudinal data which show profiles approaching asymptotes. Stat Med 26:2113–2130

Funatogawa I, Funatogawa T, Ohashi Y (2008a) A bivariate autoregressive linear mixed effects model for the analysis of longitudinal data. Stat Med 27:6367–6378

Funatogawa T, Funatogawa I, Takeuchi M (2008b) An autoregressive linear mixed effects model for the analysis of longitudinal data which include dropouts and show profiles approaching asymptotes. Stat Med 27:6351–6366

Heitjan DF (1991) Nonlinear modeling of serial immunologic data: a case study. J Am Stat Assoc 86:891–898

Henderson R, Diggle P, Dobson A (2000) Joint modelling of longitudinal measurements and event time data. Biostatistics 1:465–480

KDOQI (2007) KDOQI clinical practice guideline and clinical practice recommendations for anemia in chronic kidney disease: 2007 update of hemoglobin target. Am J Kidney Dis 50:471–530

Laird NM (1988) Missing data in longitudinal studies. Stat Med 7:305–315

Laird NM, Ware JH (1982) Random-effects models for longitudinal data. Biometrics 38:963–974

Lindsey JK (1993) Models for repeated measurements. Oxford University Press

Lipkovich I, Adams DH, Mallinckrodt C, Faries D, Baron D, Houston JP (2008) Evaluating dose response from flexible dose clinical trials. BMC Psychiatry 8:3

Little RJA, Rubin DB (1987) Statistical analysis with missing data. Wiley

Marder SR, Meibach RC (1994) Risperidone in the treatment of schizophrenia. Am J Psychiatry 151:825–835

Misumi T, Konishi S (2016) Mixed effects historical varying-coefficient model for evaluating dose-response in flexible dose trials. J R Stat Soc Ser C 65:331–344

Nasserinejad K, Rosmalen J, Kort W, Rizopoulos D, Lesaffre E (2016) Prediction of hemoglobin in blood donors using a latent class mixed-effects transition model. Stat Med 35:581–594

National Kidney Foundation (2003) K/DOQI clinical practice guidelines for bone metabolism and disease in chronic kidney disease. Am J Kidney Dis 42(4 Suppl 3):S1–S201

Xu J, Zeger SL (2001) Joint analysis of longitudinal data comprising repeated measures and times to events. Appl Stat 50:375–387

Xu XS, Yuan M, Nandy P (2012) Analysis of dose-response in flexible dose titration clinical studies. Pharm Stat 11:280–286

Chapter 4
Multivariate Autoregressive Linear Mixed Effects Models

Abstract Previous chapters discussed linear mixed effects models and autoregressive linear mixed effects models for analysis of longitudinal data. This chapter discusses multivariate extensions of these models. In longitudinal clinical studies, multivariate responses are often collected at each measurement time point from each subject. When two response variables, such as an efficacy measurement and a safety measurement are obviously correlated, there are advantages in analyzing the bivariate responses jointly. Parathyroid hormone (PTH) and serum calcium (Ca) measurements in the treatment of secondary hyperparathyroidism in chronic hemodialysis patients provide an example in which joint bivariate responses are of interest. We introduce multivariate longitudinal data and explain bivariate autoregressive linear mixed effects models in which the current responses are regressed on the previous responses of both variables, fixed effects, and random effects. The dependent bivariate responses approach equilibria, and the equilibria are modeled using fixed and random effects. These type of profiles are observed in long-term clinical studies. We also explain bivariate linear mixed effects models.

Keywords Autoregressive linear mixed effects model · Equilibrium Linear mixed effects model · Longitudinal · Multivariate

4.1 Multivariate Longitudinal Data and Vector Autoregressive Models

4.1.1 Multivariate Longitudinal Data

In the case of multivariate longitudinal data, one objective of the analysis is to gain an understanding of the relationship among the response variables. There are applications of bivariate longitudinal data in medicine: glomerular filtration rates (GFR) and inverse serum creatinine in chronic renal disease (Schluchter 1990); forced expiratory volume in 1 second (FEV_1) and functional residual capacity (FRC) in chronic obstructive pulmonary disease (COPD) (Zucker et al. 1995); blinks and heart rates

© The Author(s), under exclusive licence to Springer Nature Singapore Pte Ltd. 2018   77
I. Funatogawa and T. Funatogawa, *Longitudinal Data Analysis*, JSS Research Series
in Statistics, https://doi.org/10.1007/978-981-10-0077-5_4

to auditory stimulation in school-age boy (Liu et al. 2000); body mass index (BMI, weight/height^2) and log-transformed fasting insulin levels in non-insulin-dependent diabetes mellitus (NIDDM) (Jones 1993); CD4 and beta-2-microglobulin in AIDS (Sy et al. 1997); PTH and Ca in chronic hemodialysis patients (Funatogawa et al. 2008); and adrenocorticotropic hormone and cortisol in chronic fatigue syndrome and fibromyalgia (Liu et al. 2014). For other examples, see Shah et al. (1997), Zeger and Liang (1991), and Galecki (1994).

One analytical approach is to relate two variables through the correlation of random errors (Zeger and Liang 1991; Jones 1993; Sy et al. 1997). Another approach is to relate two variables through the correlation of random effects. Schuluchter (1990), Zucker et al. (1995), and Shah et al. (1997) were interested in the correlation of random effects in linear mixed effects models, for example, the correlation between two random slopes, that is, two linear time trends. In bivariate autoregressive linear mixed effects models (Funatogawa et al. 2008), the relationships between two responses are summarized by the autoregressive coefficients, the correlation of the random effects for baseline and equilibria, and the correlation of the random errors.

Repeated-series longitudinal data are another class of data, wherein multiple series of the same variable are measured in each subject. One example is intraocular pressures of the right and left eyes in ocular hypertension and glaucoma (Heitjan and Sharma 1997). The same variable is measured longitudinally under multiple conditions (left and right eyes).

Section 4.1.2 discusses vector autoregressive models for one subject. Section 4.2 discusses multivariate autoregressive linear mixed effects models. Section 4.3 shows an analytical example of PTH and Ca data. Section 4.4 discusses multivariate linear mixed effects models. Section 4.5 is an appendix and explains the direct product and parameter transformation.

4.1.2 Vector Autoregressive Models

This section shows how response levels change in multivariate autoregressive models. For simplicity, we consider a case in which there is only one subject and no random effect. Vector autoregressive (VAR) models are popular in time series analysis. A simple first-order bivariate VAR model is

$$
\begin{pmatrix} Y_{1,t} \\ Y_{2,t} \end{pmatrix} = \begin{pmatrix} \rho_{11} & \rho_{12} \\ \rho_{21} & \rho_{22} \end{pmatrix} \begin{pmatrix} Y_{1,t-1} \\ Y_{2,t-1} \end{pmatrix} + \begin{pmatrix} \beta_{1\,\text{int}} \\ \beta_{2\,\text{int}} \end{pmatrix} + \begin{pmatrix} \varepsilon_{1,t} \\ \varepsilon_{2,t} \end{pmatrix}, \ (t > 0).
\tag{4.1.1}
$$

$Y_{1,t}$ and $Y_{2,t}$ are the observed responses of the first and second variables at time t. $\beta_{1\,\text{int}}$ and $\beta_{2\,\text{int}}$ are the intercepts. $\varepsilon_{1,t}$ and $\varepsilon_{2,t}$ are random errors, and are usually assumed to be independent across time points. In later sections, a dependent error structure will be considered to take account of measurement errors. The baseline is not necessarily modeled explicitly in time series data analysis where the number of time points is large and stationarity is assumed. However, the baseline model is important in longitudinal data where the number of time points is limited. In the following model, the baseline is modeled separately,

$$
\left\{
\begin{aligned}
\begin{pmatrix} Y_{1,0} \\ Y_{2,0} \end{pmatrix} &= \begin{pmatrix} \beta_{1\,\text{base}} \\ \beta_{2\,\text{base}} \end{pmatrix} + \begin{pmatrix} \varepsilon_{1,0} \\ \varepsilon_{2,0} \end{pmatrix} \\
\begin{pmatrix} Y_{1,t} \\ Y_{2,t} \end{pmatrix} &= \begin{pmatrix} \rho_{11} & \rho_{12} \\ \rho_{21} & \rho_{22} \end{pmatrix} \begin{pmatrix} Y_{1,t-1} \\ Y_{2,t-1} \end{pmatrix} + \begin{pmatrix} \beta_{1\,\text{int}} \\ \beta_{2\,\text{int}} \end{pmatrix} + \begin{pmatrix} \varepsilon_{1,t} \\ \varepsilon_{2,t} \end{pmatrix}, \ (t>0)
\end{aligned}
\right. \tag{4.1.2}
$$

In univariate models in Chap. 2, $|\rho| < 1$ is the requirement for the response to approach the asymptote. The corresponding requirement in the bivariate model is that the absolute values of the roots of the determinantal equation $|\boldsymbol{\rho} - \lambda \mathbf{I}_2| = 0$ are less than one, where $\mathbf{I}_a$ means the $a \times a$ identity matrix,

$$
\boldsymbol{\rho} = \begin{pmatrix} \rho_{11} & \rho_{12} \\ \rho_{21} & \rho_{22} \end{pmatrix},
$$

$$
\begin{aligned}
|\boldsymbol{\rho} - \lambda \mathbf{I}_2| &= \begin{vmatrix} \rho_{11} - \lambda & \rho_{12} \\ \rho_{21} & \rho_{22} - \lambda \end{vmatrix} \\
&= (\rho_{11} - \lambda)(\rho_{22} - \lambda) - \rho_{12}\rho_{21} \\
&= \lambda^2 - (\rho_{11} + \rho_{22})\lambda + \rho_{11}\rho_{22} - \rho_{12}\rho_{21}.
\end{aligned} \tag{4.1.3}
$$

In this case, the elements of $\boldsymbol{\rho}^l$ tend to 0 as l becomes large (Harvey 1993). Furthermore, when $0 < \lambda < 1$, the changes in the responses are monotonic. In the following sections, we assume $0 < \lambda < 1$.

Given the existence of $(\mathbf{I}_2 - \boldsymbol{\rho})^{-1}$, it is expressed as

$$
\begin{aligned}
(\mathbf{I}_2 - \boldsymbol{\rho})^{-1} &= \left\{ \begin{pmatrix} 1 & 0 \\ 0 & 1 \end{pmatrix} - \begin{pmatrix} \rho_{11} & \rho_{12} \\ \rho_{21} & \rho_{22} \end{pmatrix} \right\}^{-1} = \begin{pmatrix} 1 - \rho_{11} & -\rho_{12} \\ -\rho_{21} & 1 - \rho_{22} \end{pmatrix}^{-1} \\
&= \frac{1}{(1-\rho_{11})(1-\rho_{22}) - \rho_{12}\rho_{21}} \begin{pmatrix} 1 - \rho_{22} & \rho_{12} \\ \rho_{21} & 1 - \rho_{11} \end{pmatrix}.
\end{aligned} \tag{4.1.4}
$$

For the existence of $(\mathbf{I}_2 - \boldsymbol{\rho})^{-1}$, $(1-\rho_{11})(1-\rho_{22}) - \rho_{12}\rho_{21}$ must not be zero.

4.2 Multivariate Autoregressive Linear Mixed Effects Models

In clinical studies, dependent bivariate continuous responses may approach equilibria over time. Autoregressive linear mixed effects models for bivariate longitudinal data in which current responses are regressed on the previous responses of both variables, fixed effects, and random effects. The equilibria are modeled using fixed and random effects. This model is a bivariate extension of the autoregressive linear mixed effects models for univariate longitudinal data given in Chap. 2.

Section 4.2.1 shows a simple example of bivariate autoregressive linear mixed effects models. Then, we introduce bivariate autoregressive linear mixed effects models. Three representations are provided: an autoregressive form and a marginal (unconditional) form in Sect. 4.2.2, and response changes with equilibria in Sect. 4.2.3. Vector representations are provided for each time point in each subject or for all time points in each subject. Section 4.2.4 provides variance covariance structures. Section 4.2.5 provides estimation methods.

4.2.1 Example of Bivariate Autoregressive Linear Mixed Effects Models

An example of a bivariate autoregressive linear mixed effects model without covariates is shown below. Let $Y_{ri,t}$ be the observed response of the rth ($r = 1, 2$) variable for the ith ($i = 1, \cdots, N$) subject at time t ($t = 0, 1, \cdots, T_i$). As with the univariate model (2.2.1) shown in Chap. 2, the models for the baseline and later time points are

$$
\left\{
\begin{aligned}
\begin{pmatrix} Y_{1i,0} \\ Y_{2i,0} \end{pmatrix} &= \begin{pmatrix} \beta_{1\,\text{base}} \\ \beta_{2\,\text{base}} \end{pmatrix} + \begin{pmatrix} b_{1\,\text{base}\,i} \\ b_{2\,\text{base}\,i} \end{pmatrix} + \begin{pmatrix} \varepsilon_{1i,0} \\ \varepsilon_{2i,0} \end{pmatrix} \\
\begin{pmatrix} Y_{1i,t} \\ Y_{2i,t} \end{pmatrix} &= \begin{pmatrix} \rho_{11} & \rho_{12} \\ \rho_{21} & \rho_{22} \end{pmatrix} \begin{pmatrix} Y_{1i,t-1} \\ Y_{2i,t-1} \end{pmatrix} + \begin{pmatrix} \beta_{1\,\text{int}} \\ \beta_{2\,\text{int}} \end{pmatrix} + \begin{pmatrix} b_{1\,\text{int}\,i} \\ b_{2\,\text{int}\,i} \end{pmatrix} + \begin{pmatrix} \varepsilon_{1i,t} \\ \varepsilon_{2i,t} \end{pmatrix}, (t > 0)
\end{aligned}
\right.
$$

$$(4.2.1)$$

where $\rho_{rr'} (r = 1, 2, r' = 1, 2)$ is an unknown regression coefficient of $Y_{ri,t}$ on $Y_{r'i,t-1}$. Not using matrices, these equations can also be expressed as

$$
\left\{
\begin{aligned}
Y_{1i,0} &= \beta_{1\,\text{base}} + b_{1\,\text{base}\,i} + \varepsilon_{1i,0} \\
Y_{2i,0} &= \beta_{2\,\text{base}} + b_{2\,\text{base}\,i} + \varepsilon_{2i,0} \\
Y_{1i,t} &= \rho_{11} Y_{1i,t-1} + \rho_{12} Y_{2i,t-1} + \beta_{1\,\text{int}} + b_{1\,\text{int}\,i} + \varepsilon_{1i,t}, (t > 0) \\
Y_{2i,t} &= \rho_{21} Y_{1i,t-1} + \rho_{22} Y_{2i,t-1} + \beta_{2\,\text{int}} + b_{2\,\text{int}\,i} + \varepsilon_{2i,t}, (t > 0)
\end{aligned}
\right.
$$

$$(4.2.2)$$

The vector representation using $\mathbf{Y}_{i,t} = \left(Y_{1i,t}, Y_{2i,t}\right)^T$ is

$$
\begin{cases}
\mathbf{Y}_{i,0} = \boldsymbol{\beta}_{\text{base}} + \mathbf{b}_{\text{base}\,i} + \boldsymbol{\varepsilon}_{i,0} \\
\mathbf{Y}_{i,t} = \boldsymbol{\rho}\mathbf{Y}_{i,t-1} + \boldsymbol{\beta}_{\text{int}} + \mathbf{b}_{\text{int}\,i} + \boldsymbol{\varepsilon}_{i,t}, \ (t > 0)
\end{cases}, \tag{4.2.3}
$$

where the vectors are defined according to the above equations. As with the univariate models, these equations are shown with the representation of response changes with equilibria, $\mathbf{Y}_{\text{equi}\,i,t} = \left(Y_{1\,\text{equi}\,i,t}, Y_{2\,\text{equi}\,i,t}\right)^T$. The equations are

$$
\begin{cases}
\mathbf{Y}_{i,0} = \boldsymbol{\beta}_{\text{base}} + \mathbf{b}_{\text{base}\,i} + \boldsymbol{\varepsilon}_{i,0} \\
\mathbf{Y}_{i,t} - \mathbf{Y}_{i,t-1} = (\mathbf{I}_2 - \boldsymbol{\rho})\left(\mathbf{Y}_{\text{equi}\,i,t} - \mathbf{Y}_{i,t-1}\right) + \boldsymbol{\varepsilon}_{i,t}, \ (t > 0) \\
\mathbf{Y}_{\text{equi}\,i,t} = (\mathbf{I}_2 - \boldsymbol{\rho})^{-1}\left(\boldsymbol{\beta}_{\text{int}} + \mathbf{b}_{\text{int}\,i}\right) \\
\qquad\quad = \boldsymbol{\beta}_{\text{int}}^* + \mathbf{b}_{\text{int}}^*
\end{cases}, \tag{4.2.4}
$$

where * (asterisk) shows the parameters for equilibria. The parameters $\boldsymbol{\beta}_{\text{int}}$ and $\mathbf{b}_{\text{int}\,i}$ are transformed into new parameters $\boldsymbol{\beta}_{\text{int}}^*$ and $\mathbf{b}_{\text{int}}^*$ for the equilibria by multiplying by $(\mathbf{I}_2 - \boldsymbol{\rho})^{-1}$. The elements of the Eq. (4.2.4) are

$$
\begin{cases}
\begin{pmatrix} Y_{1i,0} \\ Y_{2i,0} \end{pmatrix} = \begin{pmatrix} \beta_{1\,\text{base}} \\ \beta_{2\,\text{base}} \end{pmatrix} + \begin{pmatrix} b_{1\,\text{base}\,i} \\ b_{2\,\text{base}\,i} \end{pmatrix} + \begin{pmatrix} \varepsilon_{1i,0} \\ \varepsilon_{2i,0} \end{pmatrix} \\[2ex]
\begin{pmatrix} Y_{1i,t} - Y_{1i,t-1} \\ Y_{2i,t} - Y_{2i,t-1} \end{pmatrix} = \begin{pmatrix} 1 - \rho_{11} & -\rho_{12} \\ -\rho_{21} & 1 - \rho_{22} \end{pmatrix}\begin{pmatrix} Y_{1\,\text{equi}\,i,t} - Y_{1i,t-1} \\ Y_{2\,\text{equi}\,i,t} - Y_{2i,t-1} \end{pmatrix} + \begin{pmatrix} \varepsilon_{1i,t} \\ \varepsilon_{2i,t} \end{pmatrix}, (t > 0) \\[2ex]
\begin{pmatrix} Y_{1\,\text{equi}\,i,t} \\ Y_{2\,\text{equi}\,i,t} \end{pmatrix} = \begin{pmatrix} 1 - \rho_{11} & -\rho_{12} \\ -\rho_{21} & 1 - \rho_{22} \end{pmatrix}^{-1}\left\{ \begin{pmatrix} \beta_{1\,\text{int}} \\ \beta_{2\,\text{int}} \end{pmatrix} + \begin{pmatrix} b_{1\,\text{int}\,i} \\ b_{2\,\text{int}\,i} \end{pmatrix} \right\} \\[2ex]
\qquad\quad = \begin{pmatrix} \beta_{1\,\text{int}}^* + b_{1\,\text{int}\,i}^* \\ \beta_{2\,\text{int}}^* + b_{2\,\text{int}\,i}^* \end{pmatrix}
\end{cases}
$$

$$\tag{4.2.5}$$

The changes are also shown as

$$
\begin{cases}
Y_{1i,t} - Y_{1i,t-1} = (1 - \rho_{11})\left(Y_{1\,\text{equi}\,i,t} - Y_{1i,t-1}\right) - \rho_{12}\left(Y_{2\,\text{equi}\,i,t} - Y_{2i,t-1}\right) + \varepsilon_{1i,t} \\
Y_{2i,t} - Y_{2i,t-1} = -\rho_{21}\left(Y_{1\,\text{equi}\,i,t} - Y_{1i,t-1}\right) + (1 - \rho_{22})\left(Y_{2\,\text{equi}\,i,t} - Y_{2i,t-1}\right) + \varepsilon_{2i,t}
\end{cases}. \tag{4.2.6}
$$

4.2.2   Autoregressive Form and Marginal Form

First, we show the representation for each time point in each subject. Let $Y_{ri,t}$ be the observed response of the rth $(r = 1, 2)$ variable in the ith $(i = 1, \cdots, N)$ subject at time t $(t = 0, \cdots, T_i)$. The bivariate autoregressive linear mixed effects models for bivariate longitudinal data are expressed by

$$\mathbf{Y}_{i,t} = \boldsymbol{\rho} \mathbf{Y}_{i,t-1} + \mathbf{X}_{i,t} \boldsymbol{\beta} + \mathbf{Z}_{i,t} \mathbf{b}_i + \boldsymbol{\varepsilon}_{i,t}, \tag{4.2.7}$$

where $\mathbf{Y}_{i,t} = \left(Y_{1i,t}, Y_{2i,t} \right)^T$, $\boldsymbol{\beta} = \left(\boldsymbol{\beta}_1^T, \boldsymbol{\beta}_2^T \right)^T$, $\mathbf{b}_i = \left(\mathbf{b}_{1,i}^T, \mathbf{b}_{2,i}^T \right)^T$, $\boldsymbol{\varepsilon}_{i,t} = \left(\varepsilon_{1i,t}, \varepsilon_{2i,t} \right)^T$,

$$\boldsymbol{\rho} = \begin{pmatrix} \rho_{11} & \rho_{12} \\ \rho_{21} & \rho_{22} \end{pmatrix}, \mathbf{X}_{i,t} = \begin{pmatrix} \mathbf{X}_{1i,t}^T & \mathbf{0} \\ \mathbf{0} & \mathbf{X}_{2i,t}^T \end{pmatrix}^T, \mathbf{Z}_{i,t} = \begin{pmatrix} \mathbf{Z}_{1i,t}^T & \mathbf{0} \\ \mathbf{0} & \mathbf{Z}_{2i,t}^T \end{pmatrix}^T.$$

Here, $\mathbf{A}^T$ is the transpose of a matrix $\mathbf{A}$. $\rho_{rr'} (r = 1, 2, r' = 1, 2)$ is an unknown regression coefficient of $Y_{ri,t}$ on $Y_{r'i,t-1}$, $\boldsymbol{\beta}_r$ is a $p_r \times 1$ vector of unknown fixed effects parameters, $\mathbf{X}_{ri,t}$ is a known $1 \times p_r$ design matrix for fixed effects, $\mathbf{b}_{r,i}$ is a $q_r \times 1$ vector of unknown random effects parameters, $\mathbf{Z}_{ri,t}$ is a known $1 \times q_r$ design matrix for random effects, and $\varepsilon_{ri,t}$ is a random error. When t is 0, $\mathbf{Y}_{i,t-1}$ is set to $\mathbf{0}$. It is assumed that $\mathbf{b}_i$ and $\boldsymbol{\varepsilon}_i = \left(\boldsymbol{\varepsilon}_{i,0}^T, \cdots, \boldsymbol{\varepsilon}_{i,T_i}^T \right)^T$ are both independent across subjects and independently normally distributed with the mean zero and variance covariance matrices $\mathbf{G}$ and $\mathbf{R}_i$, respectively. Although we focus on bivariate responses here, it is straightforward to generalize the formulae for an arbitrary number of outcomes.

In the model (4.2.1), the parameter vector and design matrices of fixed effects are

$$\boldsymbol{\beta}_r^T = \left(\beta_{r \text{ base}} \ \beta_{r \text{ int}} \right), \boldsymbol{\beta} = \left(\beta_{1 \text{ base}} \ \beta_{1 \text{ int}} \ \beta_{2 \text{ base}} \ \beta_{2 \text{ int}} \right)^T,$$

$$\mathbf{X}_{ri,0} = \left(1 \ 0 \right), \mathbf{X}_{i,0} = \begin{pmatrix} 1 & 0 & 0 & 0 \\ 0 & 0 & 1 & 0 \end{pmatrix},$$

$$\mathbf{X}_{ri,t} = \left(0 \ 1 \right), \mathbf{X}_{i,t} = \begin{pmatrix} 0 & 1 & 0 & 0 \\ 0 & 0 & 0 & 1 \end{pmatrix}, (t > 0).$$

$\mathbf{b}_i$ and $\mathbf{Z}_{i,t}$ are defined in the same way.

Next, we show the representation for the response at all the time points in the ith subject, $\mathbf{Y}_i = \left(\mathbf{Y}_{i,0}^T, \mathbf{Y}_{i,1}^T, \cdots, \mathbf{Y}_{i,T_i}^T \right)^T$. Let $\mathbf{F}_i$ be the $(T_i + 1) \times (T_i + 1)$ square matrix in which the elements just below the diagonal are 1 and the other elements are 0 as shown in (2.3.2). Then, the vector of previous response values is

$$(\mathbf{F}_i \otimes \mathbf{I}_2) \mathbf{Y}_i = \left(0, 0, \mathbf{Y}_{i,0}^T, \mathbf{Y}_{i,1}^T, \cdots, \mathbf{Y}_{i,T_i-1}^T \right)^T,$$

where $\otimes$ means the direct product. For details of the direct product, see Appendix in Sect. 4.5.1. Table 4.1a shows $(\mathbf{F}_i \otimes \mathbf{I}_2) \mathbf{Y}_i$ for $T_i = 2$. The product of the

Table 4.1 Vector representations in autoregressive linear mixed effects models for $T_i = 2$

(a) Vector of previous responses

$$(\mathbf{F}_i \otimes \mathbf{I}_2)\mathbf{Y}_i = \left\{ \begin{pmatrix} 0\ 0\ 0 \\ 1\ 0\ 0 \\ 0\ 1\ 0 \end{pmatrix} \otimes \begin{pmatrix} 1\ 0 \\ 0\ 1 \end{pmatrix} \right\} \mathbf{Y}_i = \begin{pmatrix} 0\ 0\ 0\ 0\ 0\ 0 \\ 0\ 0\ 0\ 0\ 0\ 0 \\ 1\ 0\ 0\ 0\ 0\ 0 \\ 0\ 1\ 0\ 0\ 0\ 0 \\ 0\ 0\ 1\ 0\ 0\ 0 \\ 0\ 0\ 0\ 1\ 0\ 0 \end{pmatrix} \begin{pmatrix} Y_{1i,0} \\ Y_{2i,0} \\ Y_{1i,1} \\ Y_{2i,1} \\ Y_{1i,2} \\ Y_{2i,2} \end{pmatrix} = \begin{pmatrix} 0 \\ 0 \\ Y_{1i,0} \\ Y_{2i,0} \\ Y_{1i,1} \\ Y_{2i,1} \end{pmatrix}$$

(b) Product of matrix of autoregressive coefficients and previous response vector

$$(\mathbf{I}_{T_i+1} \otimes \boldsymbol{\rho})(\mathbf{F}_i \otimes \mathbf{I}_2)\mathbf{Y}_i = \left\{ \begin{pmatrix} 1\ 0\ 0 \\ 0\ 1\ 0 \\ 0\ 0\ 1 \end{pmatrix} \otimes \begin{pmatrix} \rho_{11}\ \rho_{12} \\ \rho_{21}\ \rho_{22} \end{pmatrix} \right\} \left\{ \begin{pmatrix} 0\ 0\ 0 \\ 1\ 0\ 0 \\ 0\ 1\ 0 \end{pmatrix} \otimes \begin{pmatrix} 1\ 0 \\ 0\ 1 \end{pmatrix} \right\} \mathbf{Y}_i$$

$$= \begin{pmatrix} \rho_{11}\ \rho_{12}\ 0\ \ 0\ \ 0\ \ 0 \\ \rho_{21}\ \rho_{22}\ 0\ \ 0\ \ 0\ \ 0 \\ 0\ \ 0\ \ \rho_{11}\ \rho_{12}\ 0\ \ 0 \\ 0\ \ 0\ \ \rho_{21}\ \rho_{22}\ 0\ \ 0 \\ 0\ \ 0\ \ 0\ \ 0\ \ \rho_{11}\ \rho_{12} \\ 0\ \ 0\ \ 0\ \ 0\ \ \rho_{21}\ \rho_{22} \end{pmatrix} \begin{pmatrix} 0 \\ 0 \\ Y_{1i,0} \\ Y_{2i,0} \\ Y_{1i,1} \\ Y_{2i,1} \end{pmatrix} = \begin{pmatrix} 0 \\ 0 \\ \rho_{11}Y_{1i,0} + \rho_{12}Y_{2i,0} \\ \rho_{21}Y_{1i,0} + \rho_{22}Y_{2i,0} \\ \rho_{11}Y_{1i,1} + \rho_{12}Y_{2i,1} \\ \rho_{21}Y_{1i,1} + \rho_{22}Y_{2i,1} \end{pmatrix}$$

$$(\mathbf{F}_i \otimes \boldsymbol{\rho})\mathbf{Y}_i = \left\{ \begin{pmatrix} 0\ 0\ 0 \\ 1\ 0\ 0 \\ 0\ 1\ 0 \end{pmatrix} \otimes \begin{pmatrix} \rho_{11}\ \rho_{12} \\ \rho_{21}\ \rho_{22} \end{pmatrix} \right\} \mathbf{Y}_i$$

$$= \begin{pmatrix} 0\ \ 0\ \ 0\ \ \ 0\ \ 0\ 0 \\ 0\ \ 0\ \ 0\ \ \ 0\ \ 0\ 0 \\ \rho_{11}\ \rho_{12}\ 0\ \ \ 0\ \ 0\ 0 \\ \rho_{21}\ \rho_{22}\ 0\ \ \ 0\ \ 0\ 0 \\ 0\ \ 0\ \ \rho_{11}\ \rho_{12}\ 0\ 0 \\ 0\ \ 0\ \ \rho_{21}\ \rho_{22}\ 0\ 0 \end{pmatrix} \begin{pmatrix} Y_{1i,0} \\ Y_{2i,0} \\ Y_{1i,1} \\ Y_{2i,1} \\ Y_{1i,2} \\ Y_{2i,2} \end{pmatrix} = \begin{pmatrix} 0 \\ 0 \\ \rho_{11}Y_{1i,0} + \rho_{12}Y_{2i,0} \\ \rho_{21}Y_{1i,0} + \rho_{22}Y_{2i,0} \\ \rho_{11}Y_{1i,1} + \rho_{12}Y_{2i,1} \\ \rho_{21}Y_{1i,1} + \rho_{22}Y_{2i,1} \end{pmatrix}$$

matrix of the autoregressive coefficients and the previous response vector is $(\mathbf{I}_{T_i+1} \otimes \boldsymbol{\rho})(\mathbf{F}_i \otimes \mathbf{I}_2)\mathbf{Y}_i$, and it equals $(\mathbf{F}_i \otimes \boldsymbol{\rho})\mathbf{Y}_i$ because

$$(\mathbf{I}_{T_i+1} \otimes \boldsymbol{\rho})(\mathbf{F}_i \otimes \mathbf{I}_2) = \mathbf{F}_i \otimes \boldsymbol{\rho}. \tag{4.2.8}$$

The derivation of this equation is provided in (4.5.3). Table 4.1b shows $(\mathbf{I}_{T_i+1} \otimes \boldsymbol{\rho})(\mathbf{F}_i \otimes \mathbf{I}_2)\mathbf{Y}_i$ and $(\mathbf{F}_i \otimes \boldsymbol{\rho})\mathbf{Y}_i$ for $T_i = 2$.

The bivariate autoregressive linear mixed effects models are expressed by

$$\mathbf{Y}_i = \left(\mathbf{I}_{T_i+1} \otimes \boldsymbol{\rho}\right)\left(\mathbf{F}_i \otimes \mathbf{I}_2\right)\mathbf{Y}_i + \mathbf{X}_i\boldsymbol{\beta} + \mathbf{Z}_i\mathbf{b}_i + \boldsymbol{\varepsilon}_i$$
$$= (\mathbf{F}_i \otimes \boldsymbol{\rho})\mathbf{Y}_i + \mathbf{X}_i\boldsymbol{\beta} + \mathbf{Z}_i\mathbf{b}_i + \boldsymbol{\varepsilon}_i, \tag{4.2.9}$$

where $\mathbf{X}_i = \left(\mathbf{X}_{i,0}^T, \cdots, \mathbf{X}_{i,T_i}^T\right)^T$, $\mathbf{Z}_i = \left(\mathbf{Z}_{i,0}^T, \cdots, \mathbf{Z}_{i,T_i}^T\right)^T$, and $\boldsymbol{\varepsilon}_i = \left(\boldsymbol{\varepsilon}_{i,0}^T, \cdots, \boldsymbol{\varepsilon}_{i,T_i}^T\right)^T$. The variance covariance matrix of $\mathbf{Y}_i$ is $\mathbf{V}_i = \mathrm{Var}(\mathbf{Z}_i\mathbf{b}_i + \boldsymbol{\varepsilon}_i) = \mathbf{Z}_i\mathbf{G}\mathbf{Z}_i^T + \mathbf{R}_i$. The model (4.2.1) for $T_i = 2$ is

$$
\begin{pmatrix} Y_{1i,0} \\ Y_{2i,0} \\ Y_{1i,1} \\ Y_{2i,1} \\ Y_{1i,2} \\ Y_{2i,2} \end{pmatrix}
=
\begin{pmatrix} 0 \\ 0 \\ \rho_{11}Y_{1i,0} + \rho_{12}Y_{2i,0} \\ \rho_{21}Y_{1i,0} + \rho_{22}Y_{2i,0} \\ \rho_{11}Y_{1i,1} + \rho_{12}Y_{2i,1} \\ \rho_{21}Y_{1i,1} + \rho_{22}Y_{2i,1} \end{pmatrix}
+
\begin{pmatrix} 1 & 0 & 0 & 0 \\ 0 & 0 & 1 & 0 \\ 0 & 1 & 0 & 0 \\ 0 & 0 & 0 & 1 \\ 0 & 1 & 0 & 0 \\ 0 & 0 & 0 & 1 \end{pmatrix}
\begin{pmatrix} \beta_{1\ \text{base}} \\ \beta_{1\ \text{int}} \\ \beta_{2\ \text{base}} \\ \beta_{2\ \text{int}} \end{pmatrix}
$$

$$
+
\begin{pmatrix} 1 & 0 & 0 & 0 \\ 0 & 0 & 1 & 0 \\ 0 & 1 & 0 & 0 \\ 0 & 0 & 0 & 1 \\ 0 & 1 & 0 & 0 \\ 0 & 0 & 0 & 1 \end{pmatrix}
\begin{pmatrix} b_{1\ \text{base}\ i} \\ b_{1\ \text{int}\ i} \\ b_{2\ \text{base}\ i} \\ b_{2\ \text{int}\ i} \end{pmatrix}
+
\begin{pmatrix} \varepsilon_{1i,0} \\ \varepsilon_{2i,0} \\ \varepsilon_{1i,1} \\ \varepsilon_{2i,1} \\ \varepsilon_{1i,2} \\ \varepsilon_{2i,2} \end{pmatrix}. \tag{4.2.10}
$$

The marginal form of Eq. (4.2.9) is

$$\mathbf{Y}_i = \left(\mathbf{I}_{2T_i+2} - \mathbf{F}_i \otimes \boldsymbol{\rho}\right)^{-1}(\mathbf{X}_i\boldsymbol{\beta} + \mathbf{Z}_i\mathbf{b}_i + \boldsymbol{\varepsilon}_i). \tag{4.2.11}$$

This equation is derived from multiplying both sides of the following equation by $\left(\mathbf{I}_{2T_i+2} - \mathbf{F}_i \otimes \boldsymbol{\rho}\right)^{-1}$,

$$\mathbf{Y}_i - (\mathbf{F}_i \otimes \boldsymbol{\rho})\mathbf{Y}_i = \mathbf{X}_i\boldsymbol{\beta} + \mathbf{Z}_i\mathbf{b}_i + \boldsymbol{\varepsilon}_i.$$

The marginal variance covariance matrix of $\mathbf{Y}_i$ is

$$\boldsymbol{\Sigma}_i = \left(\mathbf{I}_{2T_i+2} - \mathbf{F}_i \otimes \boldsymbol{\rho}\right)^{-1}\left(\mathbf{Z}_i\mathbf{G}\mathbf{Z}_i^T + \mathbf{R}_i\right)\left\{\left(\mathbf{I}_{2T_i+2} - \mathbf{F}_i \otimes \boldsymbol{\rho}\right)^{-1}\right\}^T. \tag{4.2.12}$$

The marginal form at each time point is

$$
\begin{cases}
\mathbf{Y}_{i,0} = \mathbf{X}_{i,0}\boldsymbol{\beta} + \mathbf{Z}_{i,0}\mathbf{b}_i + \boldsymbol{\varepsilon}_{i,0} \\
\mathbf{Y}_{i,t} = \sum_{l=0}^{t} \boldsymbol{\rho}^{t-l}\left(\mathbf{X}_{i,t}\boldsymbol{\beta} + \mathbf{Z}_{i,t}\mathbf{b}_i + \boldsymbol{\varepsilon}_{i,t}\right), (t > 0)
\end{cases}
\tag{4.2.13}
$$

4.2.3 Representation of Response Changes with Equilibria

Here, we consider the case that $\boldsymbol{\beta}_r^T = \left(\boldsymbol{\beta}_{r\,\text{base}}^T \;\; \boldsymbol{\beta}_{r\,\text{equi}}^T \right)$ and $\mathbf{b}_{r\,i}^T = \left(\mathbf{b}_{r\,\text{base}\,i}^T \;\; \mathbf{b}_{r\,\text{equi}\,i}^T \right)$. $\boldsymbol{\beta}_{r\,\text{base}}$ and $\mathbf{b}_{r\,\text{base}\,i}$ correspond to the baseline parts ($t = 0$), and $\boldsymbol{\beta}_{r\,\text{equi}}$ and $\mathbf{b}_{r\,\text{equi}\,i}$ correspond to the other parts ($t > 0$), and these do not overlap. Bivariate autoregressive linear mixed effects models are shown using the representation of response changes with the equilibria at each time point,

$$
\begin{cases}
\mathbf{Y}_{i,0} = \mathbf{X}_{i,0}\boldsymbol{\beta} + \mathbf{Z}_{i,0}\mathbf{b}_i + \boldsymbol{\varepsilon}_{i,0} \\
\mathbf{Y}_{i,t} - \mathbf{Y}_{i,t-1} = (\mathbf{I}_2 - \boldsymbol{\rho})\left(\mathbf{Y}_{\text{equi}\,i,t} - \mathbf{Y}_{i,t-1}\right) + \boldsymbol{\varepsilon}_{i,t}, (t > 0) \\
\mathbf{Y}_{\text{equi}\,i,t} = (\mathbf{I}_2 - \boldsymbol{\rho})^{-1}\left(\mathbf{X}_{i,t}\boldsymbol{\beta} + \mathbf{Z}_{i,t}\mathbf{b}_i\right) \\
\qquad\quad\, = \mathbf{X}_{i,t}\boldsymbol{\beta}^* + \mathbf{Z}_{i,t}\mathbf{b}_i^*
\end{cases}
\tag{4.2.14}
$$

The expected changes from $\mathbf{Y}_{i,t-1}$ to $\mathbf{Y}_{i,t}$ is $(\mathbf{I}_2 - \boldsymbol{\rho})\left(\mathbf{Y}_{\text{equi}\,i,t} - \mathbf{Y}_{i,t-1}\right)$. As with the univariate models, $\mathbf{Y}_{\text{equi}\,i,t}$ can be interpreted as the vector of the equilibria, and is not observable. $\mathbf{Y}_{\text{equi}\,i,t} - \mathbf{Y}_{i,t-1}$ is the size remaining to the equilibria, and $\boldsymbol{\rho}$ shows how the responses approach equilibria. In particular, ρ_{12} and ρ_{21} show the influences of the other response variable.

The baselines depend linearly on fixed and random effects with the coefficients $\boldsymbol{\beta}_{r\,\text{base}}$ and $\mathbf{b}_{r\,\text{base}\,i}$. The equilibria depend linearly on the fixed and random effects, with the coefficients $\boldsymbol{\beta}_{r\,\text{equi}}^*$ and $\mathbf{b}_{r\,\text{equi}\,i}^*$, where $\boldsymbol{\beta}^* = \mathbf{M}_x\boldsymbol{\beta}$ and $\mathbf{b}_i^* = \mathbf{M}_z\mathbf{b}_i$, as in the univariate case in Sect. 2.3.2. $\mathbf{M}_x$ and $\mathbf{M}_z$ are described below. If the covariate values are the same after time t, the responses would approach the equilibria gradually. If the covariate values change, the responses would move toward new equilibria. The expectations of the baselines and equilibria are $\mathbf{X}_{i,0}\boldsymbol{\beta}(= \mathbf{X}_{i,0}\boldsymbol{\beta}^*)$ and $\mathbf{X}_{i,t}\boldsymbol{\beta}^*$. The heterogeneity among subjects on the baselines and the equilibria is represented by the vector of random effects parameters $\mathbf{b}_i^* = \mathbf{M}_z\mathbf{b}_i$, which are normally distributed with the mean zero and variance covariance matrix $\mathbf{G}^* = \mathbf{M}_z\mathbf{G}\mathbf{M}_z^T$, $\mathbf{b}_i^* \sim \text{MVN}(\mathbf{0}, \mathbf{G}^*)$. The representation (4.2.14) makes the interpretation easier than the original autoregressive form (4.2.7) or marginal form (4.2.9). Biologically, $\boldsymbol{\beta}^*$, $\mathbf{b}_i^*$, and $\mathbf{G}^*$ could be interpreted more easily than $\boldsymbol{\beta}$, $\mathbf{b}_i$, and $\mathbf{G}$.

The matrices $\mathbf{M}_x$ and $\mathbf{M}_z$ are designed to change $\boldsymbol{\beta}_{r\,\text{equi}}$ and $\mathbf{b}_{r\,\text{equi}}$ to new parameters $\boldsymbol{\beta}_{r\,\text{equi}}^*$ and $\mathbf{b}_{r\,\text{equi}\,i}^*$, but not to change $\boldsymbol{\beta}_{r\,\text{base}}$ or $\mathbf{b}_{r\,\text{base}\,i}$. $\mathbf{M}_x$ can be defined as

$$
\mathbf{M}_x = \mathbf{I}_2 \otimes \mathbf{D}_{x1} + (\mathbf{I}_2 - \boldsymbol{\rho})^{-1} \otimes \mathbf{D}_{x2}.
\tag{4.2.15}
$$

$\mathbf{D}_{x1}$ is a $p_1 \times p_1$ diagonal matrix whose first $p_{1\,\text{base}}$th elements are 1, and whose other elements are 0, where p_1 is the number of fixed effects, and $p_{1\,\text{base}}$ is the number of fixed effects corresponding to the baseline. $\mathbf{D}_{x2}$ is a $p_1 \times p_1$ diagonal matrix whose first $p_{1\,\text{base}}$th elements are 0, and whose other elements are 1. $\mathbf{M}_z$ is defined similarly. Table 4.2a and b shows an example of $\mathbf{M}_x$ and $\boldsymbol{\beta}^* = \mathbf{M}_x\boldsymbol{\beta}$ in the case of three parameters of fixed effects for each response variable with $\boldsymbol{\beta}_r^{*T} = \left(\beta_{r\,\text{base}} \ \beta_{r\,\text{int}}^* \ \beta_{r\,\text{cov}}^* \right)$. The first element corresponds to the baseline, and the other two correspond to later times. Let the elements of $(\mathbf{I}_2 - \boldsymbol{\rho})^{-1}$ be

$$(\mathbf{I}_2 - \boldsymbol{\rho})^{-1} \equiv \begin{pmatrix} \rho_a & \rho_b \\ \rho_c & \rho_d \end{pmatrix}.$$

Let $\mathbf{b}_i = (b_{1\,\text{base}\,i}, b_{1\,\text{int}\,i}, b_{1\,\text{cov}\,i}, b_{2\,\text{base}\,i}, b_{2\,\text{int}\,i}, b_{2\,\text{cov}\,i})^T$. Then, $\mathbf{M}_z = \mathbf{M}_x$. Table 4.2c shows the random effects of baseline and equilibrium $\mathbf{b}_i^* = \mathbf{M}_z\mathbf{b}_i$. Table 4.2d shows an expression of $\mathbf{G}^* = \mathbf{M}_z\mathbf{G}\mathbf{M}_z^T$ which is a 6×6 variance covariance matrix of $\mathbf{b}_i^*$.

Although the representation of response changes at all the time points in each subject (2.3.13) was given for univariate cases, only the representation at each time point is given for multivariate cases in this section.

4.2.4   Variance Covariance Structures

The autoregressive form of the variance covariance matrix is $\mathbf{V}_i = \mathbf{Z}_i\mathbf{G}\mathbf{Z}_i^T + \mathbf{R}_i$, where $\mathbf{Z}_i\mathbf{G}\mathbf{Z}_i^T$ and $\mathbf{R}_i$ represent the between-subject variability induced by random effects and within-subject variability induced by random errors, respectively. We provide a particular variance covariance structure of $\mathbf{R}_i$ induced by two types of errors, an AR(1) error and a measurement error; this is a bivariate extension of the variance covariance structures for univariate models in Sect. 2.4.1. We consider the following model:

$$\begin{cases} \left(\mathbf{Y}_{i,0} - \boldsymbol{\varepsilon}_{(\text{ME})i,0}\right) = \mathbf{X}_{i,0}\boldsymbol{\beta} + \mathbf{Z}_{i,0}\mathbf{b}_i + \boldsymbol{\varepsilon}_{(\text{AR})i,0} \\ \left(\mathbf{Y}_{i,t} - \boldsymbol{\varepsilon}_{(\text{ME})i,t}\right) = \boldsymbol{\rho}\left(\mathbf{Y}_{i,t-1} - \boldsymbol{\varepsilon}_{(\text{ME})i,t-1}\right) + \mathbf{X}_{i,t}\boldsymbol{\beta} + \mathbf{Z}_{i,t}\mathbf{b}_i + \boldsymbol{\varepsilon}_{(\text{AR})i,t}, \, (t > 0) \end{cases},$$

$$(4.2.16)$$

where $\boldsymbol{\varepsilon}_{(\text{AR})i,0}$, $\boldsymbol{\varepsilon}_{(\text{AR})i,t}$ $(t > 0)$, and $\boldsymbol{\varepsilon}_{(\text{ME})i,t}$ are 2×1 random error vectors and are assumed to follow normal distributions with the mean zero and 2×2 variance covariance matrices $\mathbf{r}_{\text{AR0}}$, $\mathbf{r}_{\text{AR}}$, and $\mathbf{r}_{\text{ME}}$, respectively. $\boldsymbol{\varepsilon}_{i,t}$ in Eq. (4.2.9) corresponds to

$$\begin{cases} \boldsymbol{\varepsilon}_{i,0} = \boldsymbol{\varepsilon}_{(\text{AR})i,0} + \boldsymbol{\varepsilon}_{(\text{ME})i,0} \\ \boldsymbol{\varepsilon}_{i,t} = \boldsymbol{\varepsilon}_{(\text{AR})i,t} + \boldsymbol{\varepsilon}_{(\text{ME})i,t} - \boldsymbol{\rho}\boldsymbol{\varepsilon}_{(\text{ME})i,t-1}, \, (t > 0) \end{cases}.$$

$$(4.2.17)$$

Table 4.2 Examples of vector representations for equilibrium

(a) Matrix for parameter transformation

$$\mathbf{M}_x = \mathbf{I}_2 \otimes \begin{pmatrix} 1 & 0 & 0 \\ 0 & 0 & 0 \\ 0 & 0 & 0 \end{pmatrix} + (\mathbf{I}_2 - \boldsymbol{\rho})^{-1} \otimes \begin{pmatrix} 0 & 0 & 0 \\ 0 & 1 & 0 \\ 0 & 0 & 1 \end{pmatrix}$$

$$= \begin{pmatrix} 1 & 0 & 0 & 0 & 0 & 0 \\ 0 & 0 & 0 & 0 & 0 & 0 \\ 0 & 0 & 0 & 0 & 0 & 0 \\ 0 & 0 & 0 & 1 & 0 & 0 \\ 0 & 0 & 0 & 0 & 0 & 0 \\ 0 & 0 & 0 & 0 & 0 & 0 \end{pmatrix} + \begin{pmatrix} 0 & 0 & 0 & 0 & 0 & 0 \\ 0 & \rho_a & 0 & 0 & \rho_b & 0 \\ 0 & 0 & \rho_a & 0 & 0 & \rho_b \\ 0 & 0 & 0 & 0 & 0 & 0 \\ 0 & \rho_c & 0 & 0 & \rho_d & 0 \\ 0 & 0 & \rho_c & 0 & 0 & \rho_d \end{pmatrix} = \begin{pmatrix} 1 & 0 & 0 & 0 & 0 & 0 \\ 0 & \rho_a & 0 & 0 & \rho_b & 0 \\ 0 & 0 & \rho_a & 0 & 0 & \rho_b \\ 0 & 0 & 0 & 1 & 0 & 0 \\ 0 & \rho_c & 0 & 0 & \rho_d & 0 \\ 0 & 0 & \rho_c & 0 & 0 & \rho_d \end{pmatrix}$$

where $(\mathbf{I}_2 - \boldsymbol{\rho})^{-1} \equiv \begin{pmatrix} \rho_a & \rho_b \\ \rho_c & \rho_d \end{pmatrix}$

(b) Fixed effects of baseline and equilibrium

$$\boldsymbol{\beta}^* = \mathbf{M}_x \boldsymbol{\beta} = \begin{pmatrix} \beta_{1\ \text{base}} \\ \beta^*_{1\ \text{int}} \\ \beta^*_{1\ \text{cov}} \\ \beta_{2\ \text{base}} \\ \beta^*_{2\ \text{int}} \\ \beta^*_{2\ \text{cov}} \end{pmatrix} = \begin{pmatrix} 1 & 0 & 0 & 0 & 0 & 0 \\ 0 & \rho_a & 0 & 0 & \rho_b & 0 \\ 0 & 0 & \rho_a & 0 & 0 & \rho_b \\ 0 & 0 & 0 & 1 & 0 & 0 \\ 0 & \rho_c & 0 & 0 & \rho_d & 0 \\ 0 & 0 & \rho_c & 0 & 0 & \rho_d \end{pmatrix} \begin{pmatrix} \beta_{1\ \text{base}} \\ \beta_{1\ \text{int}} \\ \beta_{1\ \text{cov}} \\ \beta_{2\ \text{base}} \\ \beta_{2\ \text{int}} \\ \beta_{2\ \text{cov}} \end{pmatrix} = \begin{pmatrix} \beta_{1\ \text{base}} \\ \rho_a \beta_{1\ \text{int}} + \rho_b \beta_{2\ \text{int}} \\ \rho_a \beta_{1\ \text{cov}} + \rho_b \beta_{2\ \text{cov}} \\ \beta_{2\ \text{base}} \\ \rho_c \beta_{1\ \text{int}} + \rho_d \beta_{2\ \text{int}} \\ \rho_c \beta_{1\ \text{cov}} + \rho_d \beta_{2\ \text{cov}} \end{pmatrix}$$

(c) Random effects of baseline and equilibrium $\mathbf{M}_z = \mathbf{M}_x$

$$\mathbf{b}^*_i = \mathbf{M}_z \mathbf{b}_i = \begin{pmatrix} b_{1\ \text{base}\ i} \\ b^*_{1\ \text{int}\ i} \\ b^*_{1\ \text{cov}\ i} \\ b_{2\ \text{base}\ i} \\ b^*_{2\ \text{int}\ i} \\ b^*_{2\ \text{cov}\ i} \end{pmatrix} = \begin{pmatrix} b_{1\ \text{base}\ i} \\ \rho_a b_{1\ \text{int}\ i} + \rho_b b_{2\ \text{int}\ i} \\ \rho_a b_{1\ \text{cov}\ i} + \rho_b b_{2\ \text{cov}\ i} \\ b_{2\ \text{base}\ i} \\ \rho_c b_{1\ \text{int}\ i} + \rho_d b_{2\ \text{int}\ i} \\ \rho_c b_{1\ \text{cov}\ i} + \rho_d b_{2\ \text{cov}\ i} \end{pmatrix}$$

(d) Expression of variance covariance matrix of $\mathbf{b}^*_i$, $\mathbf{G}^* = \mathbf{M}_z \mathbf{G} \mathbf{M}^T_z$

$$\mathbf{G}^* = \text{Var} \begin{pmatrix} b_{1\ \text{base}\ i} \\ b^*_{1\ \text{int}\ i} \\ b^*_{1\ \text{cov}\ i} \\ b_{2\ \text{base}\ i} \\ b^*_{2\ \text{int}\ i} \\ b^*_{2\ \text{cov}\ i} \end{pmatrix} = \begin{pmatrix} \sigma^2_{1,b} & \sigma_{1b\ 1\text{int}} & \sigma_{1b\ 1c} & \sigma_{1b\ 2b} & \sigma_{1b\ 2\text{int}} & \sigma_{1b\ 2c} \\ \sigma_{1b\ 1\text{int}} & \sigma^2_{1,\text{int}} & \sigma_{1\text{int}\ 1c} & \sigma_{1\text{int}\ 2b} & \sigma_{1\text{int}\ 2\text{int}} & \sigma_{1\text{int}\ 2c} \\ \sigma_{1b\ 1c} & \sigma_{1\text{int}\ 1c} & \sigma^2_{1,c} & \sigma_{1c\ 2b} & \sigma_{1c\ 2\text{int}} & \sigma_{1c\ 2c} \\ \sigma_{1b\ 2b} & \sigma_{1\text{int}\ 2b} & \sigma_{1c\ 2b} & \sigma^2_{2,b} & \sigma_{2b\ 2\text{int}} & \sigma_{2b\ 2c} \\ \sigma_{1b\ 2\text{int}} & \sigma_{1\text{int}\ 2\text{int}} & \sigma_{1c\ 2\text{int}} & \sigma_{2b\ 2\text{int}} & \sigma^2_{2,\text{int}} & \sigma_{2\text{int}\ 2c} \\ \sigma_{1b\ 2c} & \sigma_{1\text{int}\ 2c} & \sigma_{1c\ 2c} & \sigma_{2b\ 2c} & \sigma_{2\text{int}\ 2c} & \sigma^2_{2,c} \end{pmatrix}$$

In the autoregressive form (conditional form), $\boldsymbol{\varepsilon}_{(AR)i,t}$ is the independent error across time points. In the marginal form (unconditional form), it is a serially correlated error, that is the bivariate AR(1) error with the autoregressive coefficient $\boldsymbol{\rho}$. In the marginal form, $\boldsymbol{\varepsilon}_{(ME)i,t}$ is a temporal error that is independent across time points. The autoregressive coefficients in $\boldsymbol{\rho}$ show how the responses exclusive of $\boldsymbol{\varepsilon}_{(ME)i,t}$ change.

If the model includes the random baseline effects for both variables, its 2×2 variance covariance matrix is at the same position with $\mathbf{r}_{AR0}$ in $\mathbf{V}_i$, as in the univariate case in Table 2.3c, e, and g. Thus, a constraint on $\mathbf{r}_{AR0}$ is needed. If the process of AR(1) starts long before $t = 0$, and is stationary, $\mathbf{r}_{AR0}$ is constrained to be as follows (Harvey 1993),

$$\mathbf{r}_{AR0} = \boldsymbol{\rho}\mathbf{r}_{AR0}\boldsymbol{\rho}' + \mathbf{r}_{AR}. \qquad (4.2.18)$$

However, in clinical studies, treatments start at $t = 0$ and the process may not be stationary, especially for some time after treatment initiation. Thus, the example in Sect. 4.3 uses a non-stationary process with $\mathbf{r}_{AR0} = \mathbf{0}$ $\left(\boldsymbol{\varepsilon}_{(AR)i,0} = \mathbf{0}\right)$.

4.2.5 Estimation

-2 log-likelihood $(-2ll)$ of bivariate autoregressive linear mixed effects models is given by the marginal form, based on Eq. (4.2.11) from the following multivariate normal distribution, assuming that the responses from different subjects are independent,

$$\text{MVN}\left(\left(\mathbf{I}_{2T_i+2} - \mathbf{F}_i \otimes \boldsymbol{\rho}\right)^{-1}\mathbf{X}_i\boldsymbol{\beta}, \ \boldsymbol{\Sigma}_i \right). \qquad (4.2.19)$$

It is also given by the autoregressive form, based on Eq. (4.2.9) from the following multivariate normal distribution,

$$\text{MVN}\left((\mathbf{F}_i \otimes \boldsymbol{\rho})\mathbf{Y}_i + \mathbf{X}_i\boldsymbol{\beta}, \ \mathbf{V}_i = \mathbf{Z}_i\mathbf{G}\mathbf{Z}_i^T + \mathbf{R}_i \right). \qquad (4.2.20)$$

When there are no intermittent missing responses, both methods provide $-2ll$. When there are intermittent missing responses, but corresponding elements of the covariates are known, $-2ll$ is still given by the marginal form. However, $-2ll$ using the autoregressive form cannot be calculated, because previous responses, as covariates, are missing.

When both $\boldsymbol{\rho}$ and $\mathbf{V}_i$ are known, the maximum likelihood estimators (MLEs) of the fixed effects $\hat{\boldsymbol{\beta}}$ are given directly. To obtain the maximum likelihood estimates of the variance covariance parameters and $\boldsymbol{\rho}$, we substitute $\hat{\boldsymbol{\beta}}$ in $-2ll$ and minimize the concentrated $-2ll_{\text{MLCONC}}$ using optimization methods. In Sect. 4.3, we use a Newton–Raphson ridge method with finite difference approximations for first- and second-order derivatives in the nonlinear optimization subroutines of the SAS/IML software. The standard errors of the parameter estimates are derived from the Hessian of the log-likelihood. We restrict the parameter space of the eigenvalues of the ρ to $0 < \lambda < 1$.

The above likelihoods are expressed by matrices whose sizes depend on the number of observations for a subject. When the number of time points increases, the matrices become large in multivariate longitudinal data. The state space representation and Kalman filter, which can calculate the marginal likelihood without using large matrices, are discussed in Chap. 6 (Funatogawa and Funatogawa 2008).

As with the univariate case in Sect. 2.5.3, an indirect method also exists. When there are no intermittent missing data, we can consider the autoregressive linear mixed effects models as linear mixed effects models by treating the previous responses as fixed effects,

$$\mathbf{Y}_i = \mathbf{X}_i^\# \boldsymbol{\beta}^\# + \mathbf{Z}_i \mathbf{b}_i + \boldsymbol{\varepsilon}_i, \tag{4.2.21}$$

where $\boldsymbol{\beta}^\# = \left(\boldsymbol{\beta}^T, \rho_{11}, \rho_{12}, \rho_{21}, \rho_{22}\right)^T$, $\mathbf{X}_i^\# = (\mathbf{X}_i \ \mathbf{X}_{i\rho})$,

$$
\mathbf{X}_{i\rho} =
\begin{pmatrix}
0 & 0 & 0 & 0 \\
0 & 0 & 0 & 0 \\
Y_{1i,0} & Y_{2i,0} & 0 & 0 \\
0 & 0 & Y_{1i,0} & Y_{2i,0} \\
Y_{1i,1} & Y_{2i,1} & 0 & 0 \\
0 & 0 & Y_{1i,1} & Y_{2i,1} \\
\vdots & \vdots & \vdots & \vdots \\
\vdots & \vdots & \vdots & \vdots \\
Y_{1i,T_i-1} & Y_{2i,T_i-1} & 0 & 0 \\
0 & 0 & Y_{1i,T_i-1} & Y_{2i,T_i-1}
\end{pmatrix}.
$$

Now $\mathbf{X}_i^\#$ is a $(2T_i + 2) \times (p + 4)$ matrix and $\boldsymbol{\beta}^\#$ is a $(p + 4) \times 1$ vector. Thus, when there are no intermittent missing data, we can use the estimation methods of the linear mixed effects models. The value of $-2ll$ is calculated by the autoregressive form. If it is assumed that $\mathbf{r}_{AR}$ is a diagonal matrix and $\mathbf{r}_{ME} = \mathbf{0}$, then standard software for linear mixed effects models can be used for the estimation. However, this structure is restrictive, and measurement errors are often observed in practice. If $\mathbf{r}_{ME} = \mathbf{0}$ is assumed, an EM algorithm has been proposed (Shah et al. 1997). It is practical to minimize the concentrated $-2ll$ using optimization methods, as mentioned above. This indirect method may be used to obtain the initial values for the direct method.

4.3 Example with Time-Dependent Covariate: PTH and Ca Data

This section shows an example of bivariate autoregressive linear mixed effects models with a time-dependent covariate using PTH and Ca data (Funatogawa et al. 2008). A univariate autoregressive linear mixed effects model was also adapted to the PTH data in Funatogawa et al. (2007). Changes in PTH and Ca are negatively correlated, and active vitamin D_3 causes the PTH level to decrease and the Ca level to rise. The injection of an active vitamin D_3 derivative is one therapeutic option for secondary hyperparathyroidism in hemodialysis patients. It is effective in reducing the PTH level. PTH and Ca have their respective target ranges, and it is clinically problematic if their levels are either too high or too low. In a clinical study (Kurokawa et al. 2000), an active vitamin D_3 derivative was administered three times a week for 28 weeks without control groups. PTH was measured biweekly and Ca was measured weekly. There were weekly dose modifications within individuals, taking into account both PTH and Ca levels and other medical conditions. The candidate doses were 0, 2.5, 5, 7.5, 10, 12.5, and 15 μg. A clear improvement was defined as a 50% reduction in PTH or a PTH level of less than 200 pg/mL without an associated Ca level over 11.5 mg/dL.

The profile of PTH, Ca, and dose in a typical patient is plotted in Fig. 4.1. In the patient, 10 μg of the drug was administered first, then PTH decreased and Ca increased. When the Ca level increased over 11.5 mg/dL, administration was stopped for safety reasons. After the decrease in the Ca level was confirmed, administration was restarted at a lower dose. Because there was large inter-individual variability in both the severity of the disease and sensitivity to dose modifications, it was difficult to infer the appropriate doses for each individual before treatment initiation. Therefore,

Fig. 4.1 Observed PTH level (closed circle), Ca level (open circle), and dose profiles for a typical patient. Funatogawa et al. (2008)

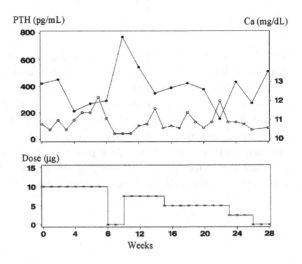

dose modifications within each individual are often needed, taking into account both PTH and Ca levels.

The objective of the analysis in this chapter is to describe the relationship between the treatment dose and the PTH and Ca levels for each patient. The number of patients was 149. The numbers of measurements was 2168 for PTH and 3910 for Ca. We use a Box–Cox transformation, $(Y^{0.25} - 1)/0.25$, for PTH to stabilize the variance and improve the model fitting.

The dose appears to follow a stochastic process, which may be interdependent with the processes for PTH and Ca. However, here, PTH and Ca are modeled conditional on the dose at a given time. If the dose is decided based on the observed levels of PTH and Ca, but not unobserved parts, and the model is specified correctly, the estimates are unbiased as discussed in Sect. 3.4 (Funatogawa and Funatogawa 2012).

Let $Y_{1i,t}$ and $Y_{2i,t}$ be PTH and Ca levels, respectively. Let $x_{i,t}$ be a treatment dose which is a time-dependent covariate. Both fixed effects and random effects include this covariate. For error terms, both first-order autoregressive errors, $\varepsilon_{(AR)i,t}$, and measurement errors, $\varepsilon_{(ME)i,t}$, are assumed. $\varepsilon_{(AR)i,t} (t > 0)$ and $\varepsilon_{(ME)i,t}$ are assumed to follow bivariate normal distributions with the mean zero vector and 2×2 variance covariance matrices $\mathbf{r}_{AR}$ and $\mathbf{r}_{ME}$, respectively. $\varepsilon_{(AR)i,0} = \mathbf{0}$ is assumed. The autoregressive form of the bivariate autoregressive linear mixed effects model is

$$
\begin{cases}
\begin{pmatrix} Y_{1i,0} \\ Y_{2i,0} \end{pmatrix} = \begin{pmatrix} \beta_{1\text{ base}} \\ \beta_{2\text{ base}} \end{pmatrix} + \begin{pmatrix} b_{1\text{ base } i} \\ b_{2\text{ base } i} \end{pmatrix} + \begin{pmatrix} \varepsilon_{1i,0} \\ \varepsilon_{2i,0} \end{pmatrix} \\
\begin{pmatrix} Y_{1i,t} \\ Y_{2i,t} \end{pmatrix} = \begin{pmatrix} \rho_{11} & \rho_{12} \\ \rho_{21} & \rho_{22} \end{pmatrix} \begin{pmatrix} Y_{1i,t-1} \\ Y_{2i,t-1} \end{pmatrix} + \begin{pmatrix} \beta_{1\text{ int}} + b_{1\text{ int } i} \\ \beta_{2\text{ int}} + b_{2\text{ int } i} \end{pmatrix} \\
\qquad + \begin{pmatrix} \beta_{1\text{ cov}} + b_{1\text{ cov } i} \\ \beta_{2\text{ cov}} + b_{2\text{ cov } i} \end{pmatrix} x_{i,t} + \begin{pmatrix} \varepsilon_{1i,t} \\ \varepsilon_{2i,t} \end{pmatrix}, (t > 0)
\end{cases}
\tag{4.3.1}
$$

The model is also shown using the response changes with equilibria,

$$
\begin{cases}
\begin{pmatrix} Y_{1i,0} \\ Y_{2i,0} \end{pmatrix} = \begin{pmatrix} \beta_{1\text{ base}} \\ \beta_{2\text{ base}} \end{pmatrix} + \begin{pmatrix} b_{1\text{ base } i} \\ b_{2\text{ base } i} \end{pmatrix} + \begin{pmatrix} \varepsilon_{1i,0} \\ \varepsilon_{2i,0} \end{pmatrix} \\
\begin{pmatrix} Y_{1i,t} - Y_{1i,t-1} \\ Y_{2i,t} - Y_{2i,t-1} \end{pmatrix} = \begin{pmatrix} 1 - \rho_{11} & -\rho_{12} \\ -\rho_{21} & 1 - \rho_{22} \end{pmatrix} \begin{pmatrix} Y_{1\text{ equi } i,t} - Y_{1i,t-1} \\ Y_{2\text{ equi } i,t} - Y_{2i,t-1} \end{pmatrix} + \begin{pmatrix} \varepsilon_{1i,t} \\ \varepsilon_{2i,t} \end{pmatrix}, (t > 0) \\
\begin{pmatrix} Y_{1\text{ equi } i,t} \\ Y_{2\text{ equi } i,t} \end{pmatrix} = \begin{pmatrix} \beta_1^*{}_{\text{int}} + b_1^*{}_{\text{int } i} \\ \beta_2^*{}_{\text{int}} + b_2^*{}_{\text{int } i} \end{pmatrix} + \begin{pmatrix} \beta_1^*{}_{\text{cov}} + b_1^*{}_{\text{cov } i} \\ \beta_2^*{}_{\text{cov}} + b_2^*{}_{\text{cov } i} \end{pmatrix} x_{i,t}
\end{cases}
$$

$$\tag{4.3.2}$$

The derivations of $\boldsymbol{\beta}^* = \mathbf{M}_x \boldsymbol{\beta}$, $\mathbf{b}_i^* = \mathbf{M}_z \mathbf{b}_i$, and $\mathbf{G}^* = \mathbf{M}_z \mathbf{G} \mathbf{M}_z^T$ are described in Sect. 4.2.3.

Table 4.3a shows the estimates of fixed effects and autoregressive coefficients of PTH and Ca data. Figure 4.2 shows the expected mean profiles of PTH and Ca

Table 4.3 Estimates of a bivariate autoregressive linear mixed effects model for PTH and Ca data

(a) Fixed effects and autoregressive coefficients

$$
\begin{cases}
Y_{1i,0} = 16.7 + b_{1\,\text{base}\,i} + \varepsilon_{1i,0} \\
Y_{2i,0} = 9.9 + b_{2\,\text{base}\,i} + \varepsilon_{2i,0} \\
Y_{1i,t} - Y_{1i,t-1} = (1 - 0.82)\left(Y_{1\,\text{equi}\,i,t} - Y_{1i,t-1}\right) - 0.13\left(Y_{2\,\text{equi}\,i,t} - Y_{2i,t-1}\right) + \varepsilon_{1i,t},\ (t > 0) \\
Y_{2i,t} - Y_{2i,t-1} = -0.005\left(Y_{1\,\text{equi}\,i,t} - Y_{1i,t-1}\right) + (1 - 0.81)\left(Y_{2\,\text{equi}\,i,t} - Y_{2i,t-1}\right) + \varepsilon_{2i,t},\ (t > 0) \\
Y_{1\,\text{equi}\,i,t} = \left(18.1 + b^*_{1\,\text{int}\,i}\right) + \left(-0.58 + b^*_{1\,\text{cov}\,i}\right)x_{i,t} \\
Y_{2\,\text{equi}\,i,t} = \left(9.7 + b^*_{2\,\text{int}\,i}\right) + \left(0.17 + b^*_{2\,\text{cov}\,i}\right)x_{i,t}
\end{cases}
$$

$$
\begin{pmatrix} \rho_{11} & \rho_{12} \\ \rho_{21} & \rho_{22} \end{pmatrix} = \begin{pmatrix} 0.82 & 0.13 \\ 0.005 & 0.81 \end{pmatrix}
$$

(b) Covariance-SD-correlation for AR(1) error $\hat{\mathbf{r}}_{AR}$ and measurement error $\hat{\mathbf{r}}_{ME}$[a]

$$
\hat{\mathbf{r}}_{AR}:\ \begin{matrix} 0.71 & -0.10 \\ (-0.77) & 0.19 \end{matrix}, \qquad \hat{\mathbf{r}}_{ME}:\ \begin{matrix} 0.80 & -0.02 \\ (-0.08) & 0.32 \end{matrix}
$$

(c) Covariance-SD-correlation for variance covariance matrix of random effects $\hat{\mathbf{G}}^*$[a]

$b_{1\,\text{base}\,i}$	3.46	10.80	0.40	−0.63	−0.15	−0.17
$b^*_{1\,\text{int}\,i}$	(0.87)	3.60	0.13	−0.25	−0.21	−0.13
$b^*_{1\,\text{cov}\,i}$	(0.34)	(0.10)	0.34	0.03	0.07	−0.02
$b_{2\,\text{base}\,i}$	(−0.26)	(−0.10)	(0.11)	0.71	0.35	0.01
$b^*_{2\,\text{int}\,i}$	(−0.06)	(−0.09)	(0.31)	(0.75)	0.67	−0.01
$b^*_{2\,\text{cov}\,i}$	(−0.52)	(−0.40)	(−0.73)	(0.15)	(−0.13)	0.09

Funatogawa et al. (2008)
[a]Covariances are above the diagonal, standard deviations are on the diagonal, and correlations are below the diagonal

when each dose is administered continuously. At baseline, the mean PTH is 16.7 (717 pg/mL) and the mean Ca is 9.9 mg/dL. For example, when a dose of 10 μg is administered continuously, the expected equilibria are 12.3 (276 pg/mL) for PTH and 11.4 mg/dL for Ca. The changes from time 0 to time 1 are

$$
\begin{cases}
Y_{1i,1} - Y_{1i,0} = (1 - 0.82)(12.3 - 16.7) - 0.13(11.4 - 9.9) = -0.99 \\
Y_{2i,1} - Y_{2i,0} = -0.005(12.3 - 16.7) + (1 - 0.81)(11.4 - 9.9) = 0.31
\end{cases}. \tag{4.3.3}
$$

The estimates of ρ_{12} and ρ_{21} show the influences of the other response variable. In this case, the low Ca level relative to the equilibrium decreases the PTH level, and the high PTH level relative to the equilibrium increases the Ca level. Nevertheless, neither estimate was significant ($P = 0.21$ and $P = 0.41$ by the Wald test).

Table 4.3b shows the covariance-SD-correlations for $\hat{\mathbf{r}}_{AR}$ and $\hat{\mathbf{r}}_{ME}$. Covariances are above the diagonal, standard deviations are on the diagonal, and correlations are below the diagonal. The correlation in $\hat{\mathbf{r}}_{AR}$ is -0.77 ($P < 0.001$ by the Wald test).

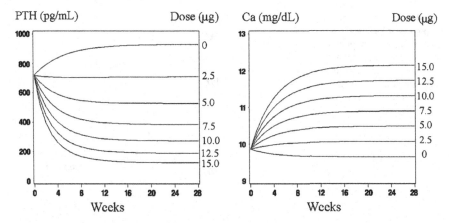

Fig. 4.2 Expected mean profiles of PTH and Ca when each dose is continuously administered. The doses are 0, 2.5, 5, 7.5, 10, 12.5, and 15 μg. Funatogawa et al. (2008)

This means that the high Ca level relative to the expected level is related to the low simultaneous PTH level relative to the expected level. The current high Ca level is also related to the high Ca levels and the low PTH levels at the neighborhood times because the errors are correlated serially. The correlation in $\hat{r}_{(ME)}$ is small ($P = 0.13$ by the Wald test), which means that the temporal errors of Ca and PTH are not correlated. Table 4.3c shows the covariance-SD-correlation for $\widehat{G}^{*}$.

We show the mean changes in the equilibria of PTH and Ca caused by 2.5 μg dose rise and the 10th and 90th percentiles. For Ca, the mean change is 0.42 (mg/dL) and the percentiles are 0.03 to 0.80. For PTH, the mean change from 700 pg/mL is -180 pg/mL and the percentiles are -310 to -10. Because a Box–Cox transformation was used for PTH, the mean change depends on the response level. These percentiles represent the heterogeneity of sensitivity to dose modification among patients. The random dose effects of PTH and Ca were negatively correlated with a correlation coefficient of -0.73 ($P < 0.001$ by the Wald test). Thus, Ca will increase markedly in patients with a large decrease in PTH, and Ca will increase by a small amount in patients with a small decrease in PTH. Figure 4.3 shows the expected dose responses of PTH and Ca equilibria for the population average and some randomly selected patients.

Table 4.4 shows the -2 log-likelihood ($-2ll$), the total number of parameters, and Akaike's information criterion (AIC) of the assumed model, the models in which r_{AR}, r_{ME}, or $\mathbf{G}$ is ignored, and the model in which ρ_{12}, ρ_{21}, and all of the correlations between PTH and Ca are set to zero. In either case, the fit is obviously worse compared with the full model. The last model corresponds to the univariate models for PTH and Ca. Thus, the bivariate analysis has an improvement over the univariate analysis.

Fig. 4.3 Expected dose responses of PTH and Ca equilibria for the population average (closed circle) and randomly selected patients (open circle). Funatogawa et al. (2008)

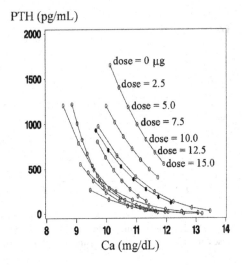

Table 4.4 -2 log-likelihood and AIC for PTH and Ca data

Model	Number of parameters	-2 log-likelihood	AIC
Full model	37	12,143.4	12,217.4
r_{AR} ignored	34	12,597.9	12,665.9
r_{ME} ignored	34	12,384.9	12,452.9
$\mathbf{G}$ ignored	16	15,403.3	15,435.3
Univariate models for PTH and Ca[a]	24	12,595.2	12,643.2

Funatogawa et al. (2008)

[a] ρ_{12}, ρ_{21}, and all of correlations between PTH and Ca are set to zero

4.4 Multivariate Linear Mixed Effects Models

The main theme of this book is autoregressive linear mixed effects models, but we also briefly describe the multivariate linear mixed effects models. Y_{rij} is the rth ($r = 1, 2$) response variable in the ith ($i = 1, \cdots, N$) subject at the jth ($j = 1, \cdots, n_{ri}$) time point. t_{rij} is time as a continuous variable. Similar to univariate models, the models can be represented as

$$\begin{cases} \mathbf{Y}_i = \mathbf{X}_i\boldsymbol{\beta} + \mathbf{Z}_i\mathbf{b}_i + \boldsymbol{\varepsilon}_i \\ \mathbf{b}_i \sim MVN(\mathbf{0}, \mathbf{G}) \\ \boldsymbol{\varepsilon}_i \sim MVN(\mathbf{0}, \mathbf{R}_i) \end{cases}, \qquad (4.4.1)$$

where $Var(\mathbf{Y}_i)$ is $\mathbf{V}_i = \mathbf{Z}_i\mathbf{G}\mathbf{Z}_i^T + \mathbf{R}_i$. When the model has no random effects, the marginal model is

Table 4.5 Example of bivariate linear mixed effects models and marginal models

(a) Example of bivariate linear mixed effects models: linear time trend model with random intercept and random slope. $\mathbf{Y}_i = \mathbf{X}_i\boldsymbol{\beta} + \mathbf{Z}_i\mathbf{b}_i + \boldsymbol{\varepsilon}_i$, $\mathbf{b}_i \sim \text{MVN}(\mathbf{0}, \mathbf{G})$, $\boldsymbol{\varepsilon}_i \sim \text{MVN}(\mathbf{0}, \mathbf{R}_i)$

$$
\begin{pmatrix} Y_{1i1} \\ \vdots \\ Y_{1in_{1i}} \\ Y_{2i1} \\ \vdots \\ Y_{2in_{2i}} \end{pmatrix} = \begin{pmatrix} 1 & 0 & t_{1i1} & 0 \\ \vdots & \vdots & \vdots & \vdots \\ 1 & 0 & t_{1in_{1i}} & 0 \\ 0 & 1 & 0 & t_{2i1} \\ \vdots & \vdots & \vdots & \vdots \\ 0 & 1 & 0 & t_{2in_{2i}} \end{pmatrix} \left\{ \begin{pmatrix} \beta_{1\,\text{int}} \\ \beta_{2\,\text{int}} \\ \beta_{1\,\text{slope}} \\ \beta_{2\,\text{slope}} \end{pmatrix} + \begin{pmatrix} b_{1\,\text{int}\,i} \\ b_{2\,\text{int}\,i} \\ b_{1\,\text{slope}\,i} \\ b_{2\,\text{slope}\,i} \end{pmatrix} \right\} + \begin{pmatrix} \varepsilon_{1i1} \\ \vdots \\ \varepsilon_{1in_{1i}} \\ \varepsilon_{2i1} \\ \vdots \\ \varepsilon_{2in_{2i}} \end{pmatrix}
$$

$$
\mathbf{G} = \begin{pmatrix} \sigma^2_{1\text{int}} & \sigma_{1\text{int}\,2\text{int}} & \sigma_{1\text{int}\,1s} & \sigma_{1\text{int}\,2s} \\ \sigma_{1\text{int}\,2\text{int}} & \sigma^2_{2\text{int}} & \sigma_{2\text{int}\,1s} & \sigma_{2\text{int}\,2s} \\ \sigma_{1\text{int}\,1s} & \sigma_{2\text{int}\,1s} & \sigma^2_{1s} & \sigma_{1s\,2s} \\ \sigma_{1\text{int}\,2s} & \sigma_{2\text{int}\,2s} & \sigma_{1s\,2s} & \sigma^2_{2s} \end{pmatrix}, \quad \mathbf{R}_i = \begin{pmatrix} \sigma^2_1\mathbf{I}_{n_{1i}} & 0 \\ 0 & \sigma^2_2\mathbf{I}_{n_{2i}} \end{pmatrix}
$$

(b) Example of variance covariance matrix of bivariate marginal model for three time points: direct product of unstructured (UN) and AR(1). $\mathbf{Y}_i = \mathbf{X}_i\boldsymbol{\beta} + \boldsymbol{\varepsilon}_i$, $\boldsymbol{\varepsilon}_i \sim \text{MVN}(\mathbf{0}, \mathbf{R}_i)$

$$
\mathbf{R}_i = \begin{pmatrix} \sigma^2_1 & \sigma_{12} \\ \sigma_{12} & \sigma^2_2 \end{pmatrix} \otimes \begin{pmatrix} 1 & \rho & \rho^2 \\ \rho & 1 & \rho \\ \rho^2 & \rho & 1 \end{pmatrix} = \begin{pmatrix} \sigma^2_1 & \sigma^2_1\rho & \sigma^2_1\rho^2 & \sigma_{12} & \sigma_{12}\rho & \sigma_{12}\rho^2 \\ \sigma^2_1\rho & \sigma^2_1 & \sigma^2_1\rho & \sigma_{12}\rho & \sigma_{12} & \sigma_{12}\rho \\ \sigma^2_1\rho^2 & \sigma^2_1\rho & \sigma^2_1 & \sigma_{12}\rho^2 & \sigma_{12}\rho & \sigma_{12} \\ \sigma_{12} & \sigma_{12}\rho & \sigma_{12}\rho^2 & \sigma^2_2 & \sigma^2_2\rho & \sigma^2_2\rho^2 \\ \sigma_{12}\rho & \sigma_{12} & \sigma_{12}\rho & \sigma^2_2\rho & \sigma^2_2 & \sigma^2_2\rho \\ \sigma_{12}\rho^2 & \sigma_{12}\rho & \sigma_{12} & \sigma^2_2\rho^2 & \sigma^2_2\rho & \sigma^2_2 \end{pmatrix}
$$

$$
\begin{cases} \mathbf{Y}_i = \mathbf{X}_i\boldsymbol{\beta} + \boldsymbol{\varepsilon}_i \\ \boldsymbol{\varepsilon}_i \sim \text{MVN}(\mathbf{0}, \mathbf{R}_i) \end{cases}, \tag{4.4.2}
$$

where $\text{Var}(\mathbf{Y}_i)$ is $\mathbf{R}_i$.

Table 4.5a shows an example of bivariate linear mixed effects models, assuming a linear time trend for each response variable. The random effects are random intercepts and random slopes for the two variables. They follow a multivariate normal distribution, with the mean vector $\mathbf{0}$ and 4×4 variance covariance matrix $\mathbf{G}$. The structure of $\mathbf{G}$ is assumed to be unstructured (UN). $\mathbf{G}$ is given in Table 4.5a. $\sigma_{1s\,2s}$ is the covariance for the two random slopes $b_{1\,\text{slope}\,i}$ and $b_{2\,\text{slope}\,i}$, and its correlation is $\sigma_{1s\,2s}/(\sigma_{1s}\sigma_{2s})$. The large absolute value of the correlation means that a subject with large changes in one variable shows large changes in the other variable. If we assume independent errors, the structure of $\mathbf{R}_i$ is diagonal. $\mathbf{R}_i$ is given in Table 4.5a. Correlated error structures can be also assumed.

In another approach with error terms, some structures are assumed on $\mathbf{R}_i$. Jones (1993) assumed a bivariate continuous time AR(1) error structure with a random intercept for each variable. Sy et al. (1997) assumed a bivariate integrated Ornstein–Uhlenbeck (IOU) process with measurement errors. Zeger and Liang (1991) were interested in the feedback of one response on another. With continuous responses, they used a bivariate AR(1) process for the error, but not for the response variable. Besides these, there is an approach using a direct (Kronecker) product. An example of the direct product of unstructured (UN) and AR(1) for the three time points is given in Table 4.5b.

4.5   Appendix

4.5.1   Direct Product

The direct product of two matrices, an $a_1 \times a_2$ matrix $\mathbf{A}_{a_1 \times a_2}$ and a $b_1 \times b_2$ matrix $\mathbf{B}_{b_1 \times b_2}$, is defined as

$$
\mathbf{A}_{a_1 \times a_2} \otimes \mathbf{B}_{b_1 \times b_2} = \begin{pmatrix} a_{11}\mathbf{B} & \cdots & a_{1a_2}\mathbf{B} \\ \vdots & & \vdots \\ a_{a_1 1}\mathbf{B} & \cdots & a_{a_1 a_2}\mathbf{B} \end{pmatrix}, \tag{4.5.1}
$$

where a_{kl} is the k, lth element of $\mathbf{A}_{a_1 \times a_2}$. This is also called the Kronecker product. The size of this matrix is $a_1 b_1 \times a_2 b_2$. Direct products have many useful properties. One property used in this chapter is

$$
(\mathbf{A} \otimes \mathbf{B})(\mathbf{C} \otimes \mathbf{D}) = \mathbf{AC} \otimes \mathbf{BD}, \tag{4.5.2}
$$

provided conformability requirements for regular matrix multiplication are satisfied. We can obtain the following equation based on (4.5.2),

$$
\begin{aligned}
\left(\mathbf{I}_{T_i+1} \otimes \boldsymbol{\rho}\right)(\mathbf{F}_i \otimes \mathbf{I}_2) &= \mathbf{I}_{T_i+1}\mathbf{F}_i \otimes \boldsymbol{\rho}\mathbf{I}_2 \\
&= \mathbf{F}_i \otimes \boldsymbol{\rho}. \tag{4.5.3}
\end{aligned}
$$

For details of the matrix algebra, see Searle (1982).

4.5.2   Parameter Transformation

This section gives a supplementary explanation of the parameter transformation in Sect. 4.2.3. When $\mathbf{X}_{1i,t} = \mathbf{X}_{2i,t}$, $\mathbf{X}_{i,t} = \mathbf{I}_2 \otimes \mathbf{X}_{1i,t}$. We can obtain

$$(\mathbf{I}_2 - \boldsymbol{\rho})^{-1}\mathbf{X}_{i,t}\boldsymbol{\beta} = \mathbf{X}_{i,t}\big\{(\mathbf{I}_2 - \boldsymbol{\rho})^{-1} \otimes \mathbf{I}_p\big\}\boldsymbol{\beta}, \qquad (4.5.4)$$

from the following equation:

$$\begin{aligned}
(\mathbf{I}_2 - \boldsymbol{\rho})^{-1}\mathbf{X}_{i,t} &= (\mathbf{I}_2 - \boldsymbol{\rho})^{-1}\big(\mathbf{I}_2 \otimes \mathbf{X}_{1i,t}\big) \\
&= (\mathbf{I}_2 - \boldsymbol{\rho})^{-1} \otimes \mathbf{X}_{1i,t} \\
&= \mathbf{I}_2(\mathbf{I}_2 - \boldsymbol{\rho})^{-1} \otimes \mathbf{X}_{1i,t}\mathbf{I}_p \\
&= \big(\mathbf{I}_2 \otimes \mathbf{X}_{1i,t}\big)\big\{(\mathbf{I}_2 - \boldsymbol{\rho})^{-1} \otimes \mathbf{I}_p\big\} \\
&= \mathbf{X}_{i,t}\big\{(\mathbf{I}_2 - \boldsymbol{\rho})^{-1} \otimes \mathbf{I}_p\big\}. \qquad (4.5.5)
\end{aligned}$$

To obtain this equation, we use the property (4.5.2). In the above derivation, only the parts of $\big\{(\mathbf{I}_2 - \boldsymbol{\rho})^{-1} \otimes \mathbf{I}_p\big\}\boldsymbol{\beta}$ corresponding to $\mathbf{X}_{1i,t,\text{equi}}^T$ and $\mathbf{X}_{2i,t,\text{equi}}^T$ are used as coefficients for the equilibria, because all elements in $\mathbf{X}_{i,t}^T$ for $t > 0$ are zero except $\mathbf{X}_{1i,t,\text{equi}}^T$ and $\mathbf{X}_{2i,t,\text{equi}}^T$.

When $\mathbf{X}_{1i,t} \neq \mathbf{X}_{2i,t}$ or $\mathbf{Z}_{1i,t} \neq \mathbf{Z}_{2i,t}$, the equilibrium $\mathbf{Y}_{\text{equi}\,i,t}$ can be also defined as a linear function of covariates. When $\mathbf{X}_{1i,t} \neq \mathbf{X}_{2i,t}$, $\mathbf{X}_{i,t}$ is represented as $\big(\mathbf{I}_2 \otimes \mathbf{x}_{i,t}^*\big)\mathbf{J}_x$. $\mathbf{x}_{i,t}^*$ is a $1 \times p^*$ ($p^* \geq p_1$ and $p^* \geq p_2$) matrix, which contains all covariates of fixed effects. $\mathbf{J}_x$ is a $2p^* \times (p_1 + p_2)$ matrix, and the elements of which are 0 or 1. For random effects, $\mathbf{z}_{i,t}^*$, $\mathbf{J}_z$, and q^* are defined similarly. We can obtain

$$(\mathbf{I}_2 - \boldsymbol{\rho})^{-1}\mathbf{X}_{i,t}\boldsymbol{\beta} = \big(\mathbf{I}_2 \otimes \mathbf{x}_{i,t}^*\big)\big\{(\mathbf{I}_2 - \boldsymbol{\rho})^{-1} \otimes \mathbf{I}_{p*}\big\}\mathbf{J}_x\boldsymbol{\beta}, \qquad (4.5.6)$$

$$(\mathbf{I}_2 - \boldsymbol{\rho})^{-1}\mathbf{Z}_{i,t}\mathbf{b}_i = \big(\mathbf{I}_2 \otimes \mathbf{z}_{i,t}^*\big)\big\{(\mathbf{I}_2 - \boldsymbol{\rho})^{-1} \otimes \mathbf{I}_{q*}\big\}\mathbf{J}_z\mathbf{b}_i. \qquad (4.5.7)$$

Thus, the equilibrium depends linearly on both the fixed and random effects with the covariate matrices, $\big(\mathbf{I}_2 \otimes \mathbf{x}_{i,t}^*\big)$ and $\big(\mathbf{I}_2 \otimes \mathbf{z}_{i,t}^*\big)$, and the parameter vectors, $\big\{(\mathbf{I}_2 - \boldsymbol{\rho})^{-1} \otimes \mathbf{I}_{p*}\big\}\mathbf{J}_x\boldsymbol{\beta}$ and $\big\{(\mathbf{I}_2 - \boldsymbol{\rho})^{-1} \otimes \mathbf{I}_{q*}\big\}\mathbf{J}_z\mathbf{b}_i$.

References

Funatogawa I, Funatogawa T (2008) State space representation of an autoregressive linear mixed effects model for the analysis of longitudinal data. In: JSM Proceedings, biometrics section. American Statistical Association, pp 3057–3062

Funatogawa I, Funatogawa T (2012) Dose-response relationship from longitudinal data with response-dependent dose-modification using likelihood methods. Biometrical J 54:494–506

Funatogawa I, Funatogawa T, Ohashi Y (2007) An autoregressive linear mixed effects model for the analysis of longitudinal data which show profiles approaching asymptotes. Stat Med 26:2113–2130

Funatogawa I, Funatogawa T, Ohashi Y (2008) A bivariate autoregressive linear mixed effects model for the analysis of longitudinal data. Stat Med 27:6367–6378

Galecki AT (1994) General class of covariance structures for two or more repeated factors in longitudinal data analysis. Commun Stat Theor Methods 23:3105–3109

Harvey AC (1993) Time series models, 2nd edn. The MIT Press

Heitjan DF, Sharma D (1997) Modelling repeated-series longitudinal data. Stat Med 16:347–355

Jones RH (1993) Longitudinal data with serial correlation: a state-space approach. Chapman & Hall

Kurokawa K, Akizawa T, Suzuki M, Akiba T, Nishizawa Y, Ohashi Y, Ogata E, Slatopolsky E (2000) Effect of long-term administration of 22-oxacalcitriol (OCT) on secondary hyperparathyroidism in hemodialysis patients. Kidney Dial 48:875–897 (in Japanese)

Liu M, Taylor JMG, Belin TR (2000) Multiple imputation and posterior simulation for multivariate missing data in longitudinal studies. Biometrics 56:1157–1163

Liu Z, Cappola AR, Crofford LJ, Guo W (2014) Modeling bivariate longitudinal hormone profiles by hierarchical state space models. J Am Stat Assoc 109:108–118

Schuluchter MD (1990) Estimating correlation between alternative measures of disease progression in a longitudinal study. Stat Med 9:1175–1188

Searl SR (1982) Matrix algebra useful for statistics. Wiley

Shah A, Laird N, Schoenfeld D (1997) A random-effects models for multiple characteristics with possibly missing data. J Am Stat Assoc 92:775–779

Sy JP, Taylor JMG, Cumberland WG (1997) A stochastic model for the analysis of bivariate longitudinal AIDS data. Biometrics 53:542–555

Zeger SL, Liang K-Y (1991) Feedback models for discrete and continuous time series. Statistica Sinica 1:51–64

Zucker DM, Zerbe GO, Wu MC (1995) Inference for the association between coefficients in a multivariate growth curve model. Biometrics 51:413–424

Chapter 5
Nonlinear Mixed Effects Models, Growth Curves, and Autoregressive Linear Mixed Effects Models

Abstract In the previous chapters, we discussed autoregressive linear mixed effects models. In this section, we discuss the relationships between the autoregressive linear mixed effects models and nonlinear mixed effects models, growth curves, and differential equations. The autoregressive model shows a profile approaching an asymptote, where the change is proportional to the distance remaining to the asymptote. Autoregressive models in discrete time correspond to monomolecular curves in continuous time. Autoregressive linear mixed effects models correspond to monomolecular curves with random effects in the baseline and asymptote, and special error terms. The autoregressive coefficient is a nonlinear parameter, but all random effects parameters in the model are linear. Therefore, autoregressive linear mixed effects models are nonlinear mixed effects models without nonlinear random effects and have a closed form of likelihood. When there are time-dependent covariates, autoregressive linear mixed effects models are represented by a differential equation and random effects. The monomolecular curve is one of the popular growth curves. We introduce other growth curves, such as the logistic curves and von Bertalanffy curves, and generalizations of growth curves. Re-parameterization is often performed in nonlinear models, and various representations of re-parameterization in monomolecular and other curves are provided herein.

Keywords Autoregressive linear mixed effects model · Growth curve
Longitudinal · Monomolecular curve · Nonlinear mixed effects model

5.1 Autoregressive Models and Monomolecular Curves

This section discusses autoregressive models and monomolecular curves in a single subject. These have neither random effects nor error terms. In Sect. 5.2, we consider monomolecular curves with random effects and autoregressive linear mixed effects models for longitudinal data. Section 5.3 discusses nonlinear mixed effects models. Section 5.4 introduces various nonlinear curves. Section 5.5 discusses generalization of nonlinear curves.

© The Author(s), under exclusive licence to Springer Nature Singapore Pte Ltd. 2018   99
I. Funatogawa and T. Funatogawa, *Longitudinal Data Analysis*, JSS Research Series
in Statistics, https://doi.org/10.1007/978-981-10-0077-5_5

In autoregressive models, the change is proportional to the distance remaining to the asymptote when the autoregressive coefficient is $0 < \rho < 1$. For example, for the following autoregressive model

$$Y_t = \beta_{\text{int}} + \rho Y_{t-1}, \qquad (5.1.1)$$

the change is

$$Y_t - Y_{t-1} = (1 - \rho)\left(\frac{\beta_{\text{int}}}{1 - \rho} - Y_{t-1}\right), \qquad (5.1.2)$$

where error terms are omitted for simplicity, Y_t is the response at time t $(t = 1, \cdots, T)$, β_{int} is the intercept, $\beta_{\text{int}}/(1 - \rho) \equiv \beta_{\text{asy}}$ is the asymptote, and $\beta_{\text{int}}/(1 - \rho) - Y_{t-1}$ is the distance remaining to the asymptote. Time is discrete in autoregressive models. As described in Sect. 2.1.2, with the baseline response β_{base}, the marginal form of the above autoregressive model is $Y_0 = \beta_{\text{base}}$ and

$$Y_t = \rho^t \beta_{\text{base}} + \sum_{l=1}^{t} \rho^{t-l} \beta_{\text{int}}$$
$$= \beta_{\text{base}} \rho^t + \beta_{\text{asy}}\left(1 - \rho^t\right), \, (t > 0). \qquad (5.1.3)$$

Next, we consider monomolecular curves in which the change is proportional to the distance remaining to the asymptote based on continuous time. Let α be the asymptote as $y(x \to \infty) = \alpha$. y is the response level and x is continuous time. With $\kappa > 0$ as a proportional constant, changes in the response level based on continuous time are expressed by an ordinary differential equation,

$$\frac{dy}{dx} = \kappa(\alpha - y), \qquad (5.1.4)$$

where $\alpha - y$ is the distance remaining to the asymptote. Let α_0 be the baseline response as $y(0) = \alpha_0$. The general solution to this differential equation is

$$y(x) = \alpha - (\alpha - \alpha_0)e^{-\kappa x}. \qquad (5.1.5)$$

This is also expressed as follows:

$$y(x) = \alpha_0 e^{-\kappa x} + \alpha\left(1 - e^{-\kappa x}\right). \qquad (5.1.6)$$

This form is used in Sect. 5.2. The model has three parameters, and the correspondence with (5.1.3) is $\alpha_0 = \beta_{\text{base}}$, $\alpha = \beta_{\text{asy}}$, and $e^{-\kappa} = \rho$. The difference between (5.1.3) and (5.1.6) is that x is continuous in time, whereas t is discrete.

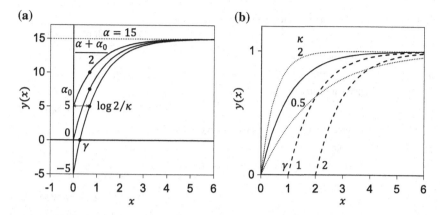

Fig. 5.1 Monomolecular curves **a** $y(x) = \alpha - (\alpha - \alpha_0)e^{-\kappa x}$ with various α_0. The curves are also expressed by $\alpha\{1 - e^{-\kappa(x-\gamma)}\}$. **b** Standard curve $y(x) = S(x) = 1 - e^{-x}$ (solid line), $S(\kappa x)$ with various κ (dotted lines), and $S(x - \gamma)$ with various γ (broken lines)

Figure 5.1a shows the monomolecular curves with various α_0. When the curve exhibits growth (increase), the parameters are $\alpha > \alpha_0 > 0$. This function can also show a decrease. κ is a scale parameter on x. When $x = \log 2/\kappa \approx 0.69314/\kappa$, the response level is the midpoint between α and α_0 as

$$y\left(\frac{\log 2}{\kappa}\right) = \frac{\alpha + \alpha_0}{2}. \tag{5.1.7}$$

With re-parameterization, the general solution can be written in many ways. For example,

$$y(x) = \alpha\{1 - e^{-\kappa(x-\gamma)}\}, \tag{5.1.8}$$

where $y(\gamma) = 0$ and

$$\gamma = \kappa^{-1}\log\left(\frac{\alpha - \alpha_0}{\alpha}\right). \tag{5.1.9}$$

Table 5.1 shows the relationships of parameters α, α_0, and κ with other parameters. Table 5.2 shows ordinary differential equations and several general solutions for the monomolecular curve and other nonlinear curves discussed in Sect. 5.4.

Let $S(x) = (1 - e^{-x})$ be a standard curve. The curve (5.1.5) is then obtained by shifting the standard curve vertically by α_0 and including the scale parameter κ and asymptote α as

$$y(x) = \alpha_0 + (\alpha - \alpha_0)S(\kappa x)$$
$$= \alpha - (\alpha - \alpha_0)\{1 - S(\kappa x)\}. \tag{5.1.10}$$

Table 5.1 Parameters of monomolecular curves

Parameter[a]		Relationship with α, α_0, κ
α	Asymptote, $y(\infty)$	
α_0	Baseline response, $y(0)$	
κ	Scale parameter on x	
γ	$y(\gamma) = 0$	$\gamma = \kappa^{-1}\log\{(\alpha - \alpha_0)/\alpha\}$
δ	Distance from baseline response to asymptote	$\delta = \alpha - \alpha_0$
β_0	Asymptote	$\beta_0 = \alpha$
β_1	Distance from asymptote to baseline response	$\beta_0 + \beta_1 = \alpha_0$ $\beta_1 = -(\alpha - \alpha_0) = -\delta$
ρ		$\rho = e^{-\kappa}, \kappa = -\log(\rho)$

[a]Monomolecular curves are expressed by $\alpha - (\alpha - \alpha_0)e^{-\kappa x}$, $\alpha\{1 - e^{-\kappa(x-\gamma)}\}$, $\alpha - \delta e^{-\kappa x}$, and $\beta_0 + \beta_1\rho^x$ with $0 < \rho < 1$

On the other hand, the curve (5.1.8) is obtained by shifting the standard curve horizontally by γ and including the scale parameter κ and asymptote α as

$$y(x) = \alpha S(\kappa(x - \gamma)). \tag{5.1.11}$$

Figure 5.1b shows $S(x)$, $S(\kappa x)$ with various κ, and $S(\kappa - \gamma)$ with various γ.
   Another re-parameterization is

$$y(x) = \alpha - \delta e^{-\kappa x}, \tag{5.1.12}$$

where $\delta = \alpha - \alpha_0$ is the distance between the baseline response and the asymptote. If we use ρ instead of $e^{-\kappa}$, the curve is

$$y(x) = \beta_0 + \beta_1\rho^x, 0 < \rho < 1, \tag{5.1.13}$$

where $\beta_0 = \alpha$, $\beta_1 = -\delta$, and $\beta_0 + \beta_1 = \alpha - \delta = \alpha_0$. When $x \geq 0$, the response changes from α_0 to α, from $\alpha - \delta$ to α, or from $\beta_0 + \beta_1$ to β_0. Other re-parameterizations are

$$y(x) = \alpha(1 - Be^{-\kappa x}), \tag{5.1.14}$$

$$y(x) = \alpha - e^{-(b+\kappa x)}. \tag{5.1.15}$$

   This monomolecular curve is also known as Mitscherlich curve, asymptotic regression (Ratkowsky 1983; Pinheiro and Bates 2000), asymptotic exponential growth curve (Vonesh 2012), negative exponential curve (Singer and Willett 2003), and so on. The form (5.1.13) or $y(x) = \beta_0 - \beta_1\rho^x$ is known as an asymptotic regression model (Ratkowsky 1983; Pinheiro and Bates 2000). The differential equation (5.1.4) is known as Newton's law of cooling, describing a body cooling over

Table 5.2 Ordinary differential equations and solutions for nonlinear curves

Nonlinear curves Ordinary differential equations	Solutions
(a) Monomolecular/Mitscherlich $\dfrac{dy}{dx} = \kappa(\alpha - y),\ \kappa > 0$	a. $y(x) = \alpha - (\alpha - \alpha_0)e^{-\kappa x}$ b. $y(x) = \alpha_0 e^{-\kappa x} + \alpha\left(1 - e^{-\kappa x}\right)$ c. $y(x) = \alpha\left\{1 - e^{-\kappa(x-\gamma)}\right\}$ d. $y(x) = \alpha - \delta e^{-\kappa x}$ e. $y(x) = \beta_0 + \beta_1 \rho^x,\ 0 < \rho < 1$ f. $y(x) = \alpha\left\{1 - Be^{-\kappa x}\right\}$ g. $y(x) = \alpha - e^{-(b+\kappa x)}$
$(\alpha_0 = 0)$	$y(x) = \alpha\left(1 - e^{-\kappa x}\right)$
(b) Exponential $\dfrac{dy}{dx} = \kappa y$	a. $y(x) = e^{\kappa(x-\gamma)}$ b. $y(x) = \alpha_0 e^{\kappa x}$
(c) Gompertz $\dfrac{dy}{dx} = \kappa y(\log \alpha - \log y),\ \kappa > 0,\ \alpha > 0$	a. $y(x) = \alpha \exp\left\{-e^{-\kappa(x-\gamma)}\right\}$ b. $y(x) = \alpha(\alpha_0/\alpha)^{\exp(-\kappa x)}$
(d) Four-parameter logistic $\dfrac{dy}{dx} = \dfrac{\kappa}{\alpha_2 - \alpha_1}(y - \alpha_1)(\alpha_2 - y)$	a. $y(x) = \alpha_1 + (\alpha_2 - \alpha_1)\left\{1 + e^{-\kappa(x-\gamma)}\right\}^{-1}$ b. $y(x) = \alpha_2 - (\alpha_2 - \alpha_1)\left\{1 + e^{\kappa(x-\gamma)}\right\}^{-1}$
(e) Three-parameter logistic, $0 < y$, $(\alpha_1 = 0$ in (d)) $\dfrac{dy}{dx} = \dfrac{\kappa}{\alpha}y(\alpha - y) = \kappa y\left(1 - \dfrac{y}{\alpha}\right)$	a. $y(x) = \alpha\left\{1 + e^{-\kappa(x-\gamma)}\right\}^{-1}$ b. $y(x) = \alpha\left\{1 + e^{(\gamma-x)/\phi}\right\}^{-1}$ c. $y(x) = \alpha\left(1 + \psi e^{-\kappa x}\right)^{-1}$ d. $y(x) = \alpha\left\{1 + e^{-(\beta_0+\beta_1 x)}\right\}^{-1}$ e. $y(x) = \alpha e^{\beta_0+\beta_1 x}\left(1 + e^{\beta_0+\beta_1 x}\right)^{-1}$
(f) Two-parameter logistic, $0 < y < 1$, $(\alpha = 1$ in (e), $\alpha_1 = 0$ and $\alpha_2 = 1$ in (d)) $\dfrac{dy}{dx} = \kappa y(1 - y)$	a. $y(x) = \left\{1 + e^{-\kappa(x-\gamma)}\right\}^{-1}$ b. $y(x) = \left\{1 + e^{(\gamma-x)/\phi}\right\}^{-1}$ c. $y(x) = \left(1 + \psi e^{-\kappa x}\right)^{-1}$ d. $y(x) = \left\{1 + e^{-(\beta_0+\beta_1 x)}\right\}^{-1}$ e. $y(x) = e^{\beta_0+\beta_1 x}\left(1 + e^{\beta_0+\beta_1 x}\right)^{-1}$

time and stating that the rate of change in the temperature of an object is proportional
to the difference between its own temperature and the ambient temperature, i.e., the
temperature of its surroundings. When used to model crop yield in the response to
the rate of fertilizer application, it has been known as Mitscherlich's law.

When we add the constraint $\alpha_0 = 0$, the response level is

$$y(x) = \alpha\left(1 - e^{-\kappa x}\right), \tag{5.1.16}$$

which is a two-parameter model for biochemical oxygen demand (BOD).

5.2 Autoregressive Linear Mixed Effects Models and Monomolecular Curves with Random Effects

Next, we consider a nonlinear mixed effects model for longitudinal data. Y_{ij} is the jth ($j = 1, \cdots, n_i$) response for the ith ($i = 1, \cdots, N$) subject. The explanatory variable is time t_{ij}, which is a continuous variable. Here, a nonlinear function, $f\left(t_{ij}, \boldsymbol{\beta}, \mathbf{b}_i\right)$, produces a monomolecular curve with three parameters. $\boldsymbol{\beta}$ is fixed effects parameters, $\mathbf{b}_i$ is random effects parameters, and ε_{ij} is an error term. The model is

$$
\begin{cases}
Y_{ij} = f\left(t_{ij}, \boldsymbol{\beta}, \mathbf{b}_i\right) + \varepsilon_{ij} \\
f\left(t_{ij}, \boldsymbol{\beta}, \mathbf{b}_i\right) = (\beta_1 + b_{1i})e^{-\beta_3 t_{ij}} + (\beta_2 + b_{2i})\left(1 - e^{-\beta_3 t_{ij}}\right) \\
\boldsymbol{\beta} = (\beta_1, \beta_2, \beta_3)^T \\
\mathbf{b}_i = (b_{1i}, b_{2i})^T, \mathbf{b}_i \sim \text{MVN}(\mathbf{0}, \mathbf{G}) \\
\boldsymbol{\varepsilon}_i = \left(\varepsilon_{i1}, \cdots, \varepsilon_{in_i}\right)^T, \boldsymbol{\varepsilon}_i \sim \text{MVN}(\mathbf{0}, \mathbf{R}_i)
\end{cases}
\tag{5.2.1}
$$

where β_1 is a baseline value, which is the expected response at time 0, β_2 is an asymptote, which is the expected response at time ∞, and β_3 is a scale parameter of time. b_{1i} and b_{2i} are random effects and represent the inter-individual variation. $\text{MVN}(\mathbf{0}, \mathbf{A})$ is a multivariate normal distribution with the mean zero vector and variance covariance matrix $\mathbf{A}$. b_{1i} and b_{2i} are linear, and only the fixed effect parameter β_3 is nonlinear. Thus, the model is partially linear. The autoregressive linear mixed effects models shown in Chap. 2 correspond to this model if there are no other explanatory variables. The autoregressive coefficient $\rho = e^{-\beta_3}$ is a fixed effect. Furthermore, as discussed in Sect. 2.4.1, the autoregressive linear mixed effects models assume an autoregressive error plus an independent error, to take a measurement error into account,

$$
\varepsilon_{i,t} = \varepsilon_{(\text{AR})i,t} + \varepsilon_{(\text{ME})i,t} - \rho \varepsilon_{(\text{ME})i,t-1}.
\tag{5.2.2}
$$

When there is a time-dependent covariate, x_{ij}, in the asymptote in (5.2.1), the nonlinear function is

$$
f\left(t_{ij}, x_{ij}, \boldsymbol{\beta}, \mathbf{b}_i\right) = (\beta_1 + b_{1i})e^{-\beta_3 t_{ij}} + \left\{(\beta_2 + b_{2i}) + (\beta_c + b_{ci})x_{ij}\right\}\left(1 - e^{-\beta_3 t_{ij}}\right).
\tag{5.2.3}
$$

In this model, Y_{ij} is expressed by the current covariate x_{ij}, but not past covariates. In contrast, the marginal form of the autoregressive linear mixed effects model with a time-dependent covariate $x_{i,t}$ shown in Sect. 2.3.3 was

$$
Y_{i,t} = \rho^t(\beta_{\text{base}} + b_{\text{base }i}) + \sum_{l=1}^{t} \rho^{t-l}\left\{\beta_{\text{int}} + b_{\text{int }i} + (\beta_{\text{cov}} + b_{\text{cov }i})x_{i,l}\right\} + \varepsilon_{\text{m }i,t}, \tag{5.2.4}
$$

for $t > 0$. In this model, $Y_{i,t}$ is expressed by the current covariate and past covariates. Therefore, the two models are different. If the covariate is not time-dependent, the autoregressive model is

$$Y_{i,t} = \rho^t(\beta_{base} + b_{base\,i}) + \left(1 - \rho^t\right)\{\beta_{int} + b_{int\,i} + (\beta_{cov} + b_{cov\,i})x_i\} + \varepsilon_{m\,i,t}. \quad (5.2.5)$$

The two models differ only whether time is continuous or discrete.

When the covariate is time-dependent, the autoregressive linear mixed effects model corresponds to the following model with a differential equation, where time is continuous,

$$\begin{cases} \dfrac{d\mu_i(t)}{dt} = \kappa\{\beta_2 + b_{2i} + (\beta_c + b_{ci})x_i(t) - \mu_i(t)\} \\[2mm] \mu_i(0) = \beta_1 + b_{1i} \\[2mm] Y_{ij} = \mu_i\left(t_{ij}\right) + \varepsilon_{ij} \\[2mm] \mathbf{b}_i = (b_{1i}, b_{2i}, b_{ci})^T, \mathbf{b}_i \sim MVN(\mathbf{0}, \mathbf{G}) \\[2mm] \boldsymbol{\varepsilon}_i = \left(\varepsilon_{i1}, \cdots, \varepsilon_{in_i}\right)^T, \boldsymbol{\varepsilon}_i \sim MVN(\mathbf{0}, \mathbf{R}_i) \end{cases} \quad (5.2.6)$$

In the following model, the random effect is only a random intercept, b_i.

$$f\left(t_{ij}, \boldsymbol{\beta}, b_i\right) = \beta_1 e^{-\beta_3 t_{ij}} + \beta_2\left(1 - e^{-\beta_3 t_{ij}}\right) + b_i. \quad (5.2.7)$$

In this model, one fixed effect parameter, β_3, is nonlinear, but the random effect is linear. This model assumes that changes among subjects are vertically parallel, as described in Sect. 1.3.1.

5.3   Nonlinear Mixed Effects Models

5.3.1   Nonlinear Mixed Effects Models

Both fixed effects parameters $\boldsymbol{\beta}$ and random effects parameters $\mathbf{b}_i$ are linear in the following linear mixed effects models shown in Chap. 1,

$$\mathbf{Y}_i = \mathbf{X}_i\boldsymbol{\beta} + \mathbf{Z}_i\mathbf{b}_i + \boldsymbol{\varepsilon}_i. \quad (5.3.1)$$

Nonlinear mixed effects models have at least one nonlinear fixed or random effects parameter. In the case of autoregressive linear mixed effects models (Funatogawa et al. 2007, 2008a; Funatogawa et al. 2008b), all parameters are linear in the following autoregressive form in Sect. 2.3.1,

$$\mathbf{Y}_i = \rho \mathbf{F}_i \mathbf{Y}_i + \mathbf{X}_i \boldsymbol{\beta} + \mathbf{Z}_i \mathbf{b}_i + \boldsymbol{\varepsilon}_i, \tag{5.3.2}$$

where $\mathbf{F}_i$ is a square matrix whose elements just below the diagonal are 1 and the other elements are 0. However, the autoregressive parameter ρ is nonlinear in the following marginal form in Sect. 2.3.3,

$$\mathbf{Y}_i = (\mathbf{I}_i - \rho \mathbf{F}_i)^{-1} (\mathbf{X}_i \boldsymbol{\beta} + \mathbf{Z}_i \mathbf{b}_i + \boldsymbol{\varepsilon}_i), \tag{5.3.3}$$

where $\mathbf{I}_i$ is an identity matrix. Therefore, this is a nonlinear mixed effects model without nonlinear random effects parameters. ρ is a nonlinear fixed effect parameter.

In linear mixed effects models, the expectation of the response $\mathbf{Y}_i$ given random effects $\mathbf{b}_i$ for subject i is $\mathrm{E}(\mathbf{Y}|\mathbf{b}_i) = \mathbf{X}_i \boldsymbol{\beta} + \mathbf{Z}_i \mathbf{b}_i$. In nonlinear mixed effects models, the expectation of the response $\mathbf{Y}_i$ is the nonlinear function $f(\mathbf{X}_i, \mathbf{Z}_i, \boldsymbol{\beta}, \mathbf{b}_i)$, and cannot be expressed by $\mathbf{X}_i \boldsymbol{\beta} + \mathbf{Z}_i \mathbf{b}_i$, which are easier to calculate. In autoregressive linear mixed effects models, the expectation is $\mathrm{E}(\mathbf{Y}|\mathbf{b}_i) = (\mathbf{I}_i - \rho \mathbf{F}_i)^{-1} (\mathbf{X}_i \boldsymbol{\beta} + \mathbf{Z}_i \mathbf{b}_i)$ and relatively easy to calculate.

In the mixed effects approach, the expected response of a typical subject with random effects $\mathbf{0}$ is $\mathrm{E}(\mathbf{Y}|\mathbf{b} = \mathbf{0})$. The marginal expectation, $\mathrm{E}(\mathbf{Y})$, is $\mathrm{E}_\mathbf{b}\{\mathrm{E}(\mathbf{Y}|\mathbf{b})\}$, which is obtained by the integration of expectation given random effects $\mathbf{b}$, $\mathrm{E}(\mathbf{Y}|\mathbf{b})$, with respect to $\mathbf{b}$, as follows:

$$\mathrm{E}(\mathbf{Y}) = \mathrm{E}_\mathbf{b}\{\mathrm{E}(\mathbf{Y}|\mathbf{b})\} = \int \mathrm{E}(\mathbf{Y}|\mathbf{b}) \mathrm{dF}_\mathbf{b}(\mathbf{b}), \tag{5.3.4}$$

where $\mathrm{F}_\mathbf{b}(\mathbf{b})$ is the distribution function of the random effects. In linear mixed effects models and autoregressive linear mixed effects models, the expectation for a typical subject and the marginal expectation are the same,

$$\mathrm{E}(\mathbf{Y}|\mathbf{b} = \mathbf{0}) = \mathrm{E}_\mathbf{b}\{\mathrm{E}(\mathbf{Y}|\mathbf{b})\} = \mathbf{X}\boldsymbol{\beta}, \tag{5.3.5}$$

$$\mathrm{E}(\mathbf{Y}|\mathbf{b} = \mathbf{0}) = \mathrm{E}_\mathbf{b}\{\mathrm{E}(\mathbf{Y}|\mathbf{b})\} = (\mathbf{I} - \rho \mathbf{F})^{-1} \mathbf{X}\boldsymbol{\beta}. \tag{5.3.6}$$

In nonlinear mixed effects models with nonlinear random effects, however, the expectation for a typical subject and the marginal expectation are not the same,

$$f(\mathbf{X}, \mathbf{Z}, \boldsymbol{\beta}, \mathbf{b} = \mathbf{0}) \neq \mathrm{E}_\mathbf{b}\{f(\mathbf{X}, \mathbf{Z}, \boldsymbol{\beta}, \mathbf{b})\}. \tag{5.3.7}$$

Subject specific interpretation and marginal interpretation are not the same. A similar discrepancy occurs in generalized linear models, in which a nonlinear link function is used for the analysis of the discrete response variable.

In nonlinear mixed effects models, additive, exponential, or proportional errors are often used. The error term ε_{ij} is assumed to follow a normal distribution with the mean zero. The additive error is

$$Y_{ij} = f\left(t_{ij}, \boldsymbol{\beta}, \mathbf{b}_i\right) + \varepsilon_{ij}. \tag{5.3.8}$$

Response levels which are always positive, such as blood drug concentration, often show a right skewed distribution, and the following exponential error is used:

$$\log Y_{ij} = \log f\left(t_{ij}, \boldsymbol{\beta}, \mathbf{b}_i\right) + \varepsilon_{ij}, \tag{5.3.9}$$

$$Y_{ij} = f\left(t_{ij}, \boldsymbol{\beta}, \mathbf{b}_i\right)\exp\left(\varepsilon_{ij}\right), \tag{5.3.10}$$

where $\log Y_{ij}$ is expressed with an additive normal error ε_{ij}. The model assumes a log-normal distribution for Y_{ij}. When the variance of the error increases with the mean, the following proportional error is also used:

$$Y_{ij} = f\left(t_{ij}, \boldsymbol{\beta}, \mathbf{b}_i\right)\left(1 + \varepsilon_{ij}\right). \tag{5.3.11}$$

This error has a constant coefficient of variation (CV), which is the standard deviation divided by the mean. The distributions of the exponential and proportional errors are log-normal and normal, respectively. Both have constant CVs, but their shapes differ.

Random effects also sometimes exhibit right skewed distributions. In such a case, the distribution of random effects is assumed to follow a log-normal distribution,

$$\boldsymbol{\eta}_i = \exp(\boldsymbol{\beta} + \mathbf{b}_i), \tag{5.3.12}$$

$$\mathbf{b}_i \sim \text{MVN}(\mathbf{0}, \mathbf{G}). \tag{5.3.13}$$

5.3.2 Estimation

In the maximum likelihood estimation of parameters in nonlinear mixed effects models, it is common to use the marginal likelihood function, obtained by integrating the simultaneous probability density function of the response variable and the random effects with respect to the random effects. However, this function cannot usually be expressed explicitly. Additionally, when there are multiple random effects, this method becomes a multiple integration. Therefore, several approximation methods have been proposed for nonlinear random effects. In contrast, linear mixed effects models and autoregressive linear mixed effects models have closed forms of likelihood (1.5.6) and (2.5.4).

One common approach is a linear first-order approximation using Taylor expansion. The first-order method (FO method) uses a Taylor expansion around the average of the random effects (Beal and Sheiner 1982, 1988). The first-order conditional estimation method (FOCE method) uses a Taylor expansion around the Bayesian estimates of the random effects (Lindstrom and Bates 1990). These methods require only a small calculation load. However, when inter-individual variation is large, the bias of estimates based on the FO method is large, and when the number of measurements is insufficient, the FOCE method does not work well. Compared with the FO method, the FOCE method has a smaller bias but lower convergence rate in the optimization of the likelihood function, and requires more calculation. The FOCE method based on Laplace approximation has also been proposed (Wolfinger and Lin 1997).

Software for analyzing nonlinear mixed effects models is characterized by approximation methods, modeling time-dependent covariates, and the programming of differential equations. Nonlinear mixed effects models are often used in specialized fields, such as population pharmacokinetics. Software widely used in population pharmacokinetics is NONMEM (Nonlinear Mixed Effects Model), and the main approximation methods are the FO and FOCE methods. A recent version has incorporated the Markov chain Monte Carlo (MCMC) method.

When drugs are repeatedly administered to the same subject, the dose is a time-dependent covariate. Blood drug concentration depends on the current dose as well as the past dosing history. Models of blood drug concentration are more complicated than those in which responses are simply regressed on the covariates at that time. In NONMEM, programming is easy in the case of repeated administration. If there are three or more compartments or pharmacokinetics are nonlinear, then solution for the differential equations often cannot be found. In such cases, NONMEM can perform the analysis using differential equations.

The main approximation methods in the NLMIXED procedure of SAS statistical software are the numerical integration by Gauss–Hermite Quadrature, the FO method, and the FOCE method based on Laplace approximation. Programming in the form of differential equations is not possible, and programming for repeated administration is complicated. Alternative software for nonlinear mixed effects models includes the nlme package in the SPLUS software.

For more details of nonlinear models for longitudinal data analysis, see Davidian (2009), Pinheiro and Bates (2000), and Vonesh (2012).

5.4 Nonlinear Curves

The monomolecular curve is one of the popular growth curves. In this section, we introduce other nonlinear curves. The change per unit time, dy/dx, is sometimes called a change rate, as is $(dy/y)/dx$. In this section, we call dy/dx a change and $(dy/y)/dx$ a change rate. We use the following notation: $y(x)$ is a response at time x, α is an asymptote as $y(\infty) = \alpha$, α_0 is a response level at time 0 as $y(0) = \alpha_0$, κ is a constant of proportionality in a differential equation, and β_0 and β_1 are regression coefficients. Nonlinear curves are used when x is not time but another variable, such as drug dose. For more details about nonlinear curves, see Lindsey (2001), Pinheiro and Bates (2000), Ratkowsky (1983), Seber and Wild (1989), and Singer and Willett (2003).

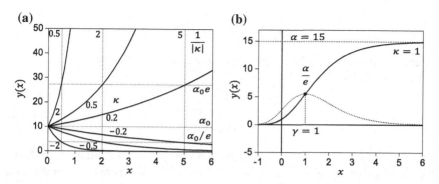

Fig. 5.2 **a** Exponential functions $y(x) = \alpha_0 e^{\kappa x}$ with various κ. **b** Gompertz curve $y(x) = \alpha \exp\{-e^{-\kappa(x-\gamma)}\}$ (solid line) with change $dy/dx = \kappa y(\log \alpha - \log y)$ (dotted line)

5.4.1 Exponential Functions

In an exponential function, the change is proportional to the current response level. With a proportional constant κ, the change is

$$\frac{dy}{dx} = \kappa y. \tag{5.4.1}$$

The change rate or growth rate is constant: $(dy/y)/dx = \kappa$. The response is expressed in several ways:

$$y(x) = e^{\kappa(x-\gamma)}, \tag{5.4.2}$$

$$y(x) = \alpha_0 e^{\kappa x}, \tag{5.4.3}$$

where $y(0) = e^{-\kappa\gamma} = \alpha_0$ is the initial response. Figure 5.2a shows the curves for various values of κ. When $\kappa > 0$, y increases, $y(1/\kappa) = \alpha_0 e$, and $y(\infty) = \infty$. The increase is initially small but becomes larger with time. When $\kappa < 0$, y decreases, $y(-1/\kappa) = \alpha_0/e$, and $y(\infty) = 0$. The decrease is initially large but becomes smaller with time. The log transformation of y makes it a linear model,

$$\log y(x) = \kappa x - \kappa\gamma, \tag{5.4.4}$$

$$\log y(x) = \kappa x + \log \alpha_0. \tag{5.4.5}$$

5.4.2 Gompertz Curves

In a Gompertz curve, the change is proportional to the current value of y and the distance remaining to the asymptote α in a log scale, $\log \alpha - \log y$. If we let κ be a proportional constant, the change is

$$\frac{dy}{dx} = \kappa y (\log \alpha - \log y),$$ (5.4.6)

where $\kappa > 0$ and $\alpha > 0$. The response is expressed in several ways:

$$y(x) = \alpha \exp\{-e^{-\kappa(x-\gamma)}\},$$ (5.4.7)

$$y(x) = \alpha \left(\frac{\alpha_0}{\alpha}\right)^{\exp(-\kappa x)},$$ (5.4.8)

where $y(-\infty) = 0$, $y(\infty) = \alpha$, and $y(0) = \alpha \exp(-e^{\kappa\gamma}) = \alpha_0$. Figure 5.2b shows the curve and dy/dx with time. γ is a point of inflection, where the change is at a maximum, and $y(\gamma) = \alpha/e$. This curve is asymmetrical about the point of inflection. After the log transformation of y, the Gompertz curve becomes a monomolecular curve. After any power transformation of y, the Gompertz curve is transformed to another Gompertz curve. This curve is used to show population growth and animal growth. Fukaya et al. (2014) applied an autoregressive model with random effects to log-transformed responses, and this model shows Gompertz curves. This is a similar approach with the autoregressive linear mixed effects models.

5.4.3 Logistic Curves

In a three-parameter logistic curve, the change is proportional to the current value of y and the distance remaining to the asymptote, $\alpha - y$. Assuming $\kappa > 0$, let $\kappa\alpha^{-1}$ be a proportional constant. The change is then

$$\frac{dy}{dx} = \frac{\kappa}{\alpha} y(\alpha - y).$$ (5.4.9)

A constant $\kappa^* = \kappa\alpha^{-1}$ is also used in $dy/dx = \kappa^* y(\alpha - y)$. Various general solutions are possible with re-parameterization. For example, the response is expressed as

$$y(x) = \frac{\alpha}{1 + e^{-\kappa(x-\gamma)}},$$ (5.4.10)

$$y(x) = \frac{\alpha}{1 + e^{(\gamma-x)/\phi}},$$ (5.4.11)

$$y(x) = \frac{\alpha}{1 + e^{-(\beta_0 + \beta_1 x)}},$$ (5.4.12)

where $e^{\kappa\gamma} = e^{\gamma/\phi} = e^{-\beta_0}$, $y(-\infty) = 0$, $y(\infty) = \alpha$, and $y(0) = \alpha/(1 + e^{\kappa\gamma})$. See Table 5.2e for other general solutions. γ is the time required for the response to become half of the asymptote, that is $y(\gamma) = \alpha/2$. γ is a point of inflection, where the change is at a maximum, and this curve is symmetric about this point. ϕ is a scale parameter as follows:

$$y(x = \gamma - \phi) = \frac{\alpha}{1+e} \approx 0.268\alpha \approx \frac{\alpha}{4}, \tag{5.4.13}$$

where ϕ is the time required for the response to change from the 1/4 asymptote to the 1/2 asymptote. When α is known, using the parameters β_0 and β_1 and the logit transformation, $\log\{y/(\alpha - y)\}$, this becomes a linear model as follows:

$$\frac{y}{\alpha - y} = \frac{\alpha}{1+e^{-(\beta_0+\beta_1 x)}} / \left\{ \frac{\alpha\left(1+e^{-(\beta_0+\beta_1 x)} - 1\right)}{1+e^{-(\beta_0+\beta_1 x)}} \right\} = \frac{1}{e^{-(\beta_0+\beta_1 x)}},$$

$$\log\left(\frac{y}{\alpha - y}\right) = \beta_0 + \beta_1 x. \tag{5.4.14}$$

In the two-parameter logistic curve with the constraint $\alpha = 1$, the change is $dy/dx = \kappa y(1-y)$. The range of the response is $0 < y < 1$. The general solutions are shown in Table 5.2f. Using the parameters β_0 and β_1 and the logit transformation, it becomes a linear model as $\log\{y/(1-y)\} = \beta_0 + \beta_1 x$.

In a four-parameter logistic curve, there are two asymptotes, α_1 and α_2, for the lower and upper bounds, respectively. Figure 5.3a shows this curve. This curve is obtained by shifting the three-parameter logistic curve by α_1 vertically. Let $S(x) = \left(1+e^{-x}\right)^{-1}$ be a standard curve. The three-parameter logistic curve is $y(x) = \alpha S(\kappa(x-\gamma))$. The four-parameter logistic curve is then

$$y(x) = \alpha_1 + (\alpha_2 - \alpha_1)S(\kappa(x-\gamma)) = \alpha_1 + \frac{\alpha_2 - \alpha_1}{1+e^{-\kappa(x-\gamma)}}, \tag{5.4.15}$$

$$y(x) = \alpha_2 - (\alpha_2 - \alpha_1)\{1 - S(\kappa(x-\gamma))\} = \alpha_2 - \frac{\alpha_2 - \alpha_1}{1+e^{\kappa(x-\gamma)}}. \tag{5.4.16}$$

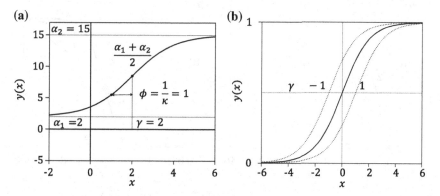

Fig. 5.3 Logistic curves **a** Four-parameter logistic curve $\alpha_1 + (\alpha_2 - \alpha_1)/\{1 + e^{-\kappa(x-\gamma)}\}$. $-\kappa(x - \gamma)$ is also expressed by $(\gamma - x)/\phi$. **b** Standard curve $y(x) = S(x) = \left(1+e^{-x}\right)^{-1}$ (solid line) and $S(x - \gamma)$ with various γ (dotted lines)

The change is shown in Table 5.2d and Fig. 5.3b shows $S(x)$ and $S(\kappa - \gamma)$ with various γ.

The probit curve under the constraint $0 < y < 1$ is the cumulative distribution function of a normal distribution. The probit curve has a shape similar to that of the logistic curve.

5.4.4 E_{max} Models and Logistic Curves

The E_{max} model, with X as an explanatory variable and the constraint $y(X = 0) = 0$, is

$$y(X) = \frac{\alpha X^\kappa}{\tau^\kappa + X^\kappa}. \tag{5.4.17}$$

Figure 5.4a shows E_{max} curves for various values of κ. In this model, $y(\infty) = \alpha$. When $y = \alpha/2$, X is τ. When the parameters are changed as $\log X = x$ and $\log \tau = \gamma$, the E_{max} model is transformed as

$$y = \frac{\alpha X^\kappa}{\tau^\kappa + X^\kappa} = \frac{\alpha}{1 + (\tau/X)^\kappa} = \frac{\alpha}{1 + e^{-\kappa(\log X - \log \tau)}} = \frac{\alpha}{1 + e^{-\kappa(x-\gamma)}}. \tag{5.4.18}$$

This is the three-parameter logistic curve (5.4.10), with x as an explanatory variable. When a common logarithm is used instead of the natural logarithm, $\log_e(10) = 2.303$ is incorporated as

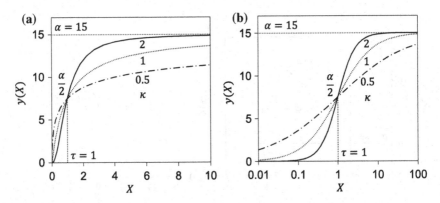

Fig. 5.4 **a** E_{max} curves $y(X) = \alpha X^\kappa/(\tau^\kappa + X^\kappa)$ with various κ. Michaelis–Menten equation when $\kappa = 1$. **b** Curves in **a** with the horizontal axis as a common logarithm. The curves are also expressed by $y(X) = \alpha/\left\{1 + e^{-2.303\kappa(\log_{10} X - \log_{10} \tau)}\right\}$ and this is three-parameter logistic curves $y(x) = \alpha/\left\{1 + e^{-\kappa(x-\gamma)}\right\}$ with $x = \log X$ and $\gamma = \log \tau$

$$y = \frac{\alpha}{1 + e^{-\kappa(\log X - \log \tau)}} = \frac{\alpha}{1 + e^{-2.303\kappa\left(\log_{10} X - \log_{10} \tau\right)}}. \tag{5.4.19}$$

The shape of the curve does not change. Figure 5.4b shows the curves in Fig. 5.4a with the horizontal axis as a common logarithm.

When $\kappa = 1$, the curve is

$$y(X) = \frac{\alpha X}{\tau + X}. \tag{5.4.20}$$

This is known as the Michaelis–Menten equation, which is used in enzyme kinetics. τ is known as the Michaelis parameter and $y(\tau) = \alpha/2$. When both the response variable and explanatory variable are transformed to their reciprocals, it becomes a linear model,

$$\frac{1}{y} = \frac{1}{\alpha} + \frac{\tau}{\alpha} \frac{1}{X}. \tag{5.4.21}$$

However, when a measurement error is the main source of errors and y follows a normal distribution, the reciprocal transformation, y^{-1}, is not adequate.

When one more parameter is added to the Michaelis–Menten equation to shift the curve vertically, the equation is

$$y(X) = \alpha_0 + \frac{(\alpha - \alpha_0)X}{\tau + X} = \alpha_0 + \frac{\alpha - \alpha_0}{\tau/X + 1} = \alpha - \frac{\alpha - \alpha_0}{\tau/X + 1} = \frac{\alpha_0 \tau + \alpha X}{\tau + X}, \tag{5.4.22}$$

where $y(X = 0) = \alpha_0$. If (α, α_0, τ) are written as (E_{max}, E_0, EC_{50}), the equation is

$$y(X) = E_0 + \frac{E_{max} - E_0}{1 + EC_{50}/X}. \tag{5.4.23}$$

When the parameter $y(X = 0) = \alpha_0$ is added to the E_{max} model with the constraint $y(X = 0) = 0$ (5.4.17) to shift the curve vertically, the equation is

$$y(X) = \frac{\alpha_0 \tau^\kappa + \alpha X^\kappa}{\tau^\kappa + X^\kappa} = \alpha - \frac{\alpha - \alpha_0}{1 + \left(\tau^{-1}X\right)^\kappa}. \tag{5.4.24}$$

This curve is known as the Morgan-Mercer-Flodin (MMF) curve.

5.4.5 Other Nonlinear Curves

The power function is $y(x) = \lambda x^\kappa$. With the proportional constant κ, the change is $dy/dx = \kappa y/x$, which is proportional to the current response level y and inversely proportional to the current time x. The elasticity, that is the ratio of the increase rate of x to that of y, is constant as $(dy/y)/(dx/x) = \kappa$. This function does not increase

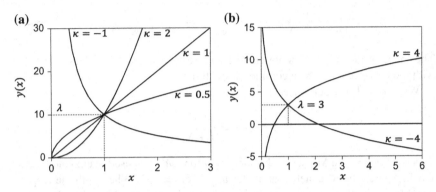

Fig. 5.5 **a** Power functions $y(x) = \lambda x^{\kappa}$ with various κ. **b** Functions $y(x) = \lambda + \kappa \log x$ with various κ

as fast as the exponential curve. Figure 5.5a shows these curves for several values of κ. In this function, $y(0) = 0$, $y(1) = \lambda$, and when $\kappa > 0$, $y(\infty) = \infty$. If both y and x are logarithmically transformed, it becomes a linear model as $\log y = \log \lambda + \kappa \log x$.

For the function $y(x) = \lambda + \kappa \log x$, the change is $dy/dx = \kappa/x$ and the elasticity is $(dy/y)/(dx/x) = \kappa/y$. At larger values of y, the elasticity becomes smaller. This function is linear in parameters and $\log x$. $y(1)$ equals λ. Figure 5.5b shows the curves for several values of κ. The function $y(x) = x/(ax - b)$ becomes a linear function with inverse transformation of y and x, as $y^{-1} = a - bx^{-1}$. The Michaelis–Menten equation (5.4.20) corresponds to this function. A rectangular hyperbola with asymptotic lines $x = b$ and $y = \alpha$ is $y = \alpha + \{\kappa(x - b)\}^{-1}$. The function $y(x) = \alpha - (\kappa x)^{-1}$ shows the changes to the asymptote α. The inverse quadratic function, $y(x) = \alpha - (\kappa_1 x + \kappa_2 x^2)^{-1}$ may also be used.

5.5 Generalization of Growth Curves

Table 5.3 shows ordinary differential equations and several general solutions for the generalization of growth curves. The von Bertalanffy curve (von Bertalanffy 1957) is

$$\frac{dy}{dx} = \eta y^{\delta} - \xi y. \tag{5.5.1}$$

The response is

$$y(x) = \left\{ \frac{\eta}{\xi} - \left(\frac{\eta}{\xi} - \alpha_0^{1-\delta} \right) e^{-(1-\delta)\xi x} \right\}^{1/(1-\delta)}, \tag{5.5.2}$$

where $y(0) = \alpha_0$. This curve is used for animal, especially fish, growth. In the narrow sense, the von Bertalanffy curve has $\delta = 2/3$ (Seber and Wild 1989).

Table 5.3 Ordinary differential equations and solutions for generalized nonlinear curves

Nonlinear curves	Ordinary differential equations Solutions
(a) von Bertalanffy	$\dfrac{dy}{dx} = \eta y^{\delta} - \xi y$ $y(x) = \left\{ \dfrac{\eta}{\xi} - \left(\dfrac{\eta}{\xi} - \alpha_0^{1-\delta} \right) e^{-(1-\delta)\xi x} \right\}^{1/(1-\delta)}$
(narrow sense, $\delta = 2/3$)	$\dfrac{dy}{dx} = \eta y^{2/3} - \xi y$ $y(x) = \alpha \left(1 + \psi e^{-\kappa x} \right)^3$
(b) Richards/generalized logistic $\delta = 0$: monomolecular $\delta = 2/3$: von Bertalanffy (narrow sense) $\delta = 2$: logistic $\delta \to 1$: Gompertz	$\dfrac{dy}{dx} = \dfrac{\kappa}{1-\delta} y \left\{ \left(\dfrac{\alpha}{y} \right)^{1-\delta} - 1 \right\}, \, \delta \neq 1$ a. $y(x) = \alpha \{ 1 + (\delta - 1)e^{-\kappa(x-\gamma)} \}^{1/(1-\delta)}$ b. $y(x) = \left[\alpha^{1-\delta} - \{ \alpha^{1-\delta} - y(x_0)^{1-\delta} \} e^{-\kappa(x-x_0)} \right]^{1/(1-\delta)}$ c. $y(x)^{1-\delta} = \alpha^{1-\delta} - \left(\alpha^{1-\delta} - \alpha_0^{1-\delta} \right) e^{-\kappa x}$ d. $y(x) = \alpha \left[1 + \{ (\alpha_0/\alpha)^{1-\delta} - 1 \} e^{-\kappa x} \right]^{1/(1-\delta)}$ e. $y(x) = \alpha \left(1 + \psi e^{-\kappa x} \right)^{-\varphi}$
(c) Modified generalized logistic $\lambda = -1$: monomolecular $\lambda = 0$: Gompertz $\lambda = 1$: logistic $\lambda \to \infty$, $g(\alpha, \lambda) \to$ const. : exponential	$\dfrac{dy}{dx} = ky\{ g(\alpha, \lambda) - g(y, \lambda) \}$ $g(y, \lambda) = \dfrac{y^{\lambda} - 1}{\lambda}, \, \lambda \neq 0$ $g(y, \lambda) = \log(y), \lambda = 0$ $y(x) = \alpha \left[1 + \{ (\alpha/\alpha_0)^{\lambda} - 1 \} e^{-k\alpha^{\lambda}x} \right]^{-1/\lambda}, \, \lambda \neq 0$ $y(x) = \alpha (\alpha_0/\alpha)^{\exp(-kx)}, \lambda = 0$

The Richards curve (Richards 1959) is

$$\frac{dy}{dx} = \frac{\kappa}{1-\delta} y \left\{ \left(\frac{\alpha}{y} \right)^{1-\delta} - 1 \right\}, \quad (\delta \neq 1). \tag{5.5.3}$$

The response is

$$y(x) = \alpha \{ 1 + (\delta - 1)e^{-\kappa(x-\gamma)} \}^{1/(1-\delta)}, \quad (\delta \neq 1). \tag{5.5.4}$$

Depending on δ, this curve includes the monomolecular curve ($\delta = 0$), the von Bertalanffy curve in the narrow sense ($\delta = 2/3$), the logistic curve ($\delta = 2$), and the Gompertz curve ($\delta \to 1$). γ is a point of inflection where the change is at a maximum, and the response level at this point is $y(\gamma) = \alpha \delta^{1/(1-\delta)}$. When $\delta \leq 0$, the curve does not have an inflection point. The response levels at the points of inflection are $8\alpha/27 \cong 0.296\alpha$ ($\delta = 2/3$), $\alpha/2 = 0.5\alpha$ ($\delta = 2$), and $\alpha/e \cong 0.368\alpha$ ($\delta \to 1$).

The Richards curve re-parameterizes the parameters (ξ, η) in the von Bertalanffy curve to (α, κ), as $\alpha^{1-\delta} = \eta/\xi$ and $\kappa = (1-\delta)\xi = (1-\delta)\eta\alpha^{-(1-\delta)}$. There are no constraints on the range of δ in the Richards curve, and $\delta > 1$ is permitted, such that it includes the logistic curve ($\delta = 2$). The Richards curve becomes a monomolecular curve when y is transformed as follows:

$$
\begin{cases}
y^{1-\delta}, & (\delta \neq 1) \\
\log(y), & (\delta \to 1)
\end{cases}
\tag{5.5.5}
$$

The curve may be written in various ways. The Richards curve with $x = x_0$ and $y(x_0)$ is

$$
y(x) = \left[\alpha^{1-\delta} - \left\{ \alpha^{1-\delta} - y(x_0)^{1-\delta} \right\} e^{-\kappa(x-x_0)} \right]^{1/(1-\delta)}.
\tag{5.5.6}
$$

With $x = 0$ and $y(0) = \alpha_0$, the curve is

$$
y(x)^{1-\delta} = \alpha^{1-\delta} - \left(\alpha^{1-\delta} - \alpha_0^{1-\delta} \right) e^{-\kappa x}.
\tag{5.5.7}
$$

Another form is shown in Table 5.3. When the parameters (δ, α_0) of the Richards curve are transformed to (φ, ψ) as $\varphi = -1/(1 - \delta)$ and $\psi = (\alpha_0/\alpha)^{1-\delta} - 1$, the curve is

$$
y(x) = \frac{\alpha}{(1 + \psi e^{-\kappa x})^{\varphi}}.
\tag{5.5.8}
$$

raising the denominator of the three-parameter logistic curve to the φth power. When $\delta = 2/3$, the von Bertalanffy curve in the narrow sense is

$$
y(x) = \alpha \left(1 + \psi e^{-\kappa x} \right)^3.
\tag{5.5.9}
$$

The modified version of the generalized logistic curve (Heitjan 1991) is

$$
\begin{cases}
\dfrac{dy}{dx} = ky\{g(\alpha, \lambda) - g(y, \lambda)\} \\
g(y, \lambda) = \dfrac{y^{\lambda} - 1}{\lambda}, \quad (\lambda \neq 0) \\
g(y, \lambda) = \log(y), \quad (\lambda = 0)
\end{cases}
\tag{5.5.10}
$$

In this case, the parameters (δ, κ) in the Richards curve are transformed to (λ, k) as $\lambda = -(1 - \delta)$ and $k = \kappa \alpha^{-\lambda}$. With $y(0) = \alpha_0$, the solution is

$$
\begin{cases}
y(x) = \alpha \left[1 + \left\{ \left(\dfrac{\alpha}{\alpha_0} \right)^{\lambda} - 1 \right\} e^{-k\alpha^{\lambda} x} \right]^{-1/\lambda}, & (\lambda \neq 0) \\
y(x) = \alpha \left(\dfrac{\alpha_0}{\alpha} \right)^{\exp(-kx)}, & (\lambda = 0)
\end{cases}
\tag{5.5.11}
$$

The curve is defined by λ and includes the monomolecular curve ($\lambda = -1$), Gompertz curve ($\lambda = 0$), logistic curve ($\lambda = 1$), and exponential curve ($\lambda \to \infty$ and $g(\alpha, \lambda) \to$ constant).

The equation $y(x) = \alpha / \left[1 + \exp\{\varphi_0 + \varphi_1 g(x)\} \right]$, where $g(x)$ is a function of x, is also called a generalized logistic curve. For $g(x)$, $\beta_1 x + \beta_2 x^2 + \beta_3 x^3$ or $\left(x^\lambda - 1\right)/\lambda$ is used. The five-parameter logistic curve is also called a generalized logistic curve.

References

Beal SL, Sheiner LB (1982) Estimating population kinetics. Crit Rev Biomed Eng 8:195–222

Beal SL, Sheiner LB (1988) Heteroscedastic nonlinear regression. Technometrics 30:327–338

Davidian M (2009) Non-linear mixed-effects models. In: Fitzmaurice G et al. (eds) Longitudinal data analysis. Chapman and Hall/CRC Series of Handbooks of modern statistical methods, pp 107–141

Fukaya K, Okuda T, Nakaoka M, Noda T (2014) Effects of spatial structure of population size on the population dynamics of barnacles across their elevational range. J Anim Ecol 83:1334–1343

Funatogawa I, Funatogawa T, Ohashi Y (2007) An autoregressive linear mixed effects model for the analysis of longitudinal data which show profiles approaching asymptotes. Stat Med 26:2113–2130

Funatogawa I, Funatogawa T, Ohashi Y (2008a) A bivariate autoregressive linear mixed effects model for the analysis of longitudinal data. Stat Med 27:6367–6378

Funatogawa T, Funatogawa I, Takeuchi M (2008b) An autoregressive linear mixed effects model for the analysis of longitudinal data which include dropouts and show profiles approaching asymptotes. Stat Med 27:6351–6366

Heitjan DF (1991) Nonlinear modeling of serial immunologic data: a case study. J Am Stat Assoc 86:891–898

Lindsey JK (2001) Nonlinear models in medical statistics. Oxford University Press

Lindstrom MJ, Bates DM (1990) Nonlinear mixed effects models for repeated measures data. Biometrics 46:673–687

Pinheiro JC, Bates DM (2000) Mixed-effects models in S and S-PLUS. Springer-Verlag

Ratkowsky DA (1983) Nonlinear Regression Modeling. Marcel Dekker

Richards FJ (1959) A flexible growth function for empirical use. J Exp Bot 10:290–300

Seber GAF, Wild GJ (1989) Nonlinear Regression. Wiley

Singer JD, Willett JB (2003) Applied longitudinal data analysis. Modeling change and event occurrence. Oxford University Press

von Bertalanffy L (1957) Quantitative laws in metabolism and growth. Q Rev Biol 32:217–231

Vonesh EF (2012) Generalized linear and nonlinear models for correlated data. Theory and applications using SAS. SAS Institute Inc

Wolfinger RD, Lin X (1997) Two Taylor-series approximation methods for nonlinear mixed models. Comput Stat Data Anal 25:465–490

Chapter 6
State Space Representations of Autoregressive Linear Mixed Effects Models

Abstract The previous chapters discussed longitudinal data analysis using linear mixed effects models and autoregressive linear mixed effects models. This chapter discusses state space representations of these models. This chapter also introduces the state space representations of time series data and the extension to multivariate longitudinal data. We use the state space representations and the Kalman filter as an alternative method to calculate the likelihoods for longitudinal data. In the autoregressive linear mixed effects models, the current response is regressed on the previous response, fixed effects, and random effects. Intermittent missing, that is the missing values in the previous response as a covariate, is an inherent problem with autoregressive models. One approach to this problem is based on the marginal form of likelihoods because they are not conditional on the previous response. State space representations with the modified Kalman filter also provide the marginal form of likelihoods without using large matrices. Calculation of likelihood usually requires matrices whose size depends on the number of observations of a subject, but this method does not. In the modified method, the regression coefficients of the fixed effects are concentrated out of the likelihood.

Keywords Autoregressive linear mixed effects model · Kalman filter Linear mixed effects model · Longitudinal · State space

6.1 Time Series Data

State space representations are often used for the analysis of time series data (Anderson and Hsiao 1982; Harvey 1993). This section discusses the state space representations of time series data before introducing the state space representations of longitudinal data in Sect. 6.2. We introduce the state space representations used in longitudinal data analysis with univariate and multivariate autoregressive linear mixed effects models in Sects. 6.3 and 6.4, respectively. We introduce the state space representations of linear mixed effects models in Sect. 6.5.

I. Funatogawa and T. Funatogawa, *Longitudinal Data Analysis*, JSS Research Series in Statistics, https://doi.org/10.1007/978-981-10-0077-5_6

6.1.1   State Space Representations of Time Series Data

Time series data are composed of multiple time points $t = 1, \cdots, T$. A state space representation consists of two equations; a state equation,

$$\mathbf{s}_{(t)} = \mathbf{\Phi}_{(t:t-1)}\mathbf{s}_{(t-1)} + \mathbf{f}_t^{(s)} + \mathbf{\upsilon}_t, \tag{6.1.1}$$

and an observation equation,

$$\mathbf{Y}_t = \mathbf{H}_t\mathbf{s}_{(t)} + \mathbf{f}_t^{(o)} + \mathbf{\xi}_t. \tag{6.1.2}$$

In the state equation, $\mathbf{s}_{(t)}$ is the state vector at time t, $\mathbf{\Phi}_{(t:t-1)}$ is the state transition matrix from $t - 1$ to t, $\mathbf{f}_t^{(s)}$ is a non-random input vector, and $\mathbf{\upsilon}_t$ is a random input vector with a variance covariance matrix $\mathbf{Q}_t \equiv \mathrm{Var}(\mathbf{\upsilon}_t)$. $\mathbf{\upsilon}_t$ and $\mathbf{Q}_t$ are also written as $\mathbf{S}\mathbf{\upsilon}_t'(= \mathbf{\upsilon}_t)$, $\mathbf{Q}_t' \equiv \mathrm{Var}(\mathbf{\upsilon}_t')$, and $\mathrm{Var}(\mathbf{S}\mathbf{\upsilon}_t') = \mathbf{S}\mathbf{Q}_t'\mathbf{S}(= \mathbf{Q}_t)$. In the observation equation, $\mathbf{Y}_t$ is the response vector at time t. In the case of univariate time series data, $\mathbf{Y}_t$ is a scalar. $\mathbf{H}_t$ indicates which elements or linear combinations of the state vector are observed, $\mathbf{f}_t^{(o)}$ is a non-random input vector, and $\mathbf{\xi}_t$ is a random input vector with a variance covariance matrix $\mathbf{r}_t \equiv \mathrm{Var}(\mathbf{\xi}_t)$. $\mathbf{f}_t^{(s)}$ and $\mathbf{f}_t^{(o)}$ are often omitted from the representations, but we follow Jones (1993) by including these. A state equation can also be called a transition equation or a system equation, and an observation equation can also be called a measurement equation.

The Kalman filter defined in the following section is a recursive algorithm that produces linear estimators of the state vector. The one-step prediction or forecast, which is the estimate of the state at time t given the observations up to time $t - 1$, is denoted by $\mathbf{s}_{(t|t-1)}$. The filter, which is the estimate of the state at time t given the observations up to time t, is denoted by $\mathbf{s}_{(t|t)}$. Smoothing, which is not covered in the following sections, is the estimate of a past state at time t ($t < T$) from all of the observations up to time T. Let the variance covariance matrices of $\mathbf{s}_{(t|t-1)}$ and $\mathbf{s}_{(t|t)}$ be $\mathbf{P}_{(t|t-1)} \equiv \mathrm{Var}(\mathbf{s}_{(t|t-1)})$ and $\mathbf{P}_{(t|t)} \equiv \mathrm{Var}(\mathbf{s}_{(t|t)})$, respectively.

The state space representation of a model is defined by the state equation with the variance covariance matrix $\mathbf{Q}_t$, the observation equation with the variance covariance matrix $\mathbf{r}_t$, and an initial state $\mathbf{s}_{(0|0)}$ with the variance covariance matrix $\mathbf{P}_{(0|0)}$, which are specified in Table 6.1a. The state space representation of a model is not necessarily unique.

We consider the state space representation of a stationary AR(1) error and a measurement error (ME). The model is

$$\begin{cases} Y_t = b + \varepsilon_{e(AR)t} + \varepsilon_{(ME)t} \\ \varepsilon_{e(AR)t} = \rho\varepsilon_{e(AR)t-1} + \eta_{(AR)t} \end{cases}, \tag{6.1.3}$$

with $\eta_{(AR)t} \sim N\left(0, \sigma_{AR}^2\right)$, $\varepsilon_{e(AR)0} \sim N\left(0, \sigma_{AR}^2\left(1 - \rho^2\right)^{-1}\right)$, and $\varepsilon_{(ME)t} \sim N\left(0, \sigma_{ME}^2\right)$, where

$$\begin{aligned} \text{Var}\left(\varepsilon_{e(AR)t}\right) &= \rho^2\text{Var}\left(\varepsilon_{e(AR)t-1}\right) + \text{Var}\left(\eta_{(AR)t}\right) \\ &= \rho^2\sigma_{AR}^2\left(1 - \rho^2\right)^{-1} + \sigma_{AR}^2 \\ &= \sigma_{AR}^2\left(1 - \rho^2\right)^{-1}. \end{aligned}$$

The subscript e in $\varepsilon_{e(AR)t}$ means an AR(1) error instead of the autoregressive model in the response. Table 6.1b shows a state space representation of this model.

6.1.2 Steps for Kalman Filter for Time Series Data

The Kalman filter calculates the following Steps 1 through 5 for each observation, starting at the first observation and repeating until the last observation. The steps, which are summarized in Table 6.2, are defined below.

Table 6.1 State space representations of time series data: (a) general and (b) with a stationary AR(1) error and a measurement error (ME)

	Equations and initial state	Variance
(a) General		
State equation	$s_{(t)} = \Phi_{(t:t-1)}s_{(t-1)} + f_t^{(s)} + \upsilon_t$	$Q_t \equiv \text{Var}(\upsilon_t)$
Observation equation	$Y_t = H_t s_{(t)} + f_t^{(o)} + \xi_t$	$r_t \equiv \text{Var}(\xi_t)$
Initial state	$s_{(0\|0)}$	$P_{(0\|0)} \equiv \text{Var}\left(s_{(0\|0)}\right)$
(b) AR(1) and ME		
State equation	$\varepsilon_{e(AR)t} = \rho\varepsilon_{e(AR)t-1} + \eta_{(AR)t}$	$\text{Var}\left(\eta_{(AR)t}\right) = \sigma_{AR}^2$
Observation equation	$Y_t = \varepsilon_{e(AR)t} + b + \varepsilon_{(ME)t}$	$\text{Var}\left(\varepsilon_{(ME)t}\right) = \sigma_{ME}^2$
Initial state	$s_{(0\|0)} = 0$	$P_{(0\|0)} = \sigma_{AR}^2\left(1 - \rho^2\right)^{-1}$

Step 1 [Prediction equations] Calculate the one-step prediction of the next state and its variance covariance matrix,

$$\mathbf{s}_{(t|t-1)} = \mathbf{\Phi}_{(t:t-1)}\mathbf{s}_{(t-1|t-1)} + \mathbf{f}_t^{(s)}, \tag{6.1.4}$$

$$\mathbf{P}_{(t|t-1)} = \mathbf{\Phi}_{(t:t-1)}\mathbf{P}_{(t-1|t-1)}\mathbf{\Phi}_{(t:t-1)}^T + \mathbf{Q}_t. \tag{6.1.5}$$

Step 2 Predict the next observation vector,

$$\mathbf{Y}_{(t|t-1)} = \mathbf{H}_t\mathbf{s}_{(t|t-1)} + \mathbf{f}_t^{(o)}, \tag{6.1.6}$$

where $\mathbf{Y}_{(t|t-1)}$ is the predicted value of $\mathbf{Y}_t$ given the observations up to time $t-1$.

Step 3 Calculate the innovation vector $\mathbf{e}_t$, which is the difference between the observation vector $\mathbf{Y}_t$ and the predicted observation vector $\mathbf{Y}_{(t|t-1)}$, and the variance covariance matrix of this innovation, $\mathbf{V}_t \equiv \mathrm{Var}(\mathbf{e}_t)$,

$$\mathbf{e}_t = \mathbf{Y}_t - \mathbf{Y}_{(t|t-1)}, \tag{6.1.7}$$

$$\mathbf{V}_t = \mathbf{H}_t\mathbf{P}_{(t|t-1)}\mathbf{H}_t^T + \mathbf{r}_t, \tag{6.1.8}$$

where $\mathbf{e}_t$ is the prediction error.

Step 4 Accumulate the following quantities, which are used to calculate -2 log-likelihood ($-2ll$),

$$M_t = M_{t-1} + \mathbf{e}_t^T\mathbf{V}_t^{-1}\mathbf{e}_t, \tag{6.1.9}$$

$$DET_t = DET_{t-1} + \log|\mathbf{V}_t|. \tag{6.1.10}$$

The initial values are $M_0 = 0$ and $DET_0 = 0$. These quantities accumulate over all of the observations.

Table 6.2 Steps for the Kalman filter

Steps	Calculation
(1) Prediction equations	$\mathbf{s}_{(t|t-1)} = \mathbf{\Phi}_{(t:t-1)}\mathbf{s}_{(t-1|t-1)} + \mathbf{f}_t^{(s)}$ $\mathbf{P}_{(t|t-1)} = \mathbf{\Phi}_{(t:t-1)}\mathbf{P}_{(t-1|t-1)}\mathbf{\Phi}_{(t:t-1)}^T + \mathbf{Q}_t$
(2) Prediction of $\mathbf{Y}_t$	$\mathbf{Y}_{(t|t-1)} = \mathbf{H}_t\mathbf{s}_{(t|t-1)} + \mathbf{f}_t^{(o)}$
(3) Innovation	$\mathbf{e}_t = \mathbf{Y}_t - \mathbf{Y}_{(t|t-1)}$ $\mathbf{V}_t = \mathbf{H}_t\mathbf{P}_{(t|t-1)}\mathbf{H}_t^T + \mathbf{r}_t$
(4) Accumulation	$M_t = M_{t-1} + \mathbf{e}_t^T\mathbf{V}_t^{-1}\mathbf{e}_t$ $DET_t = DET_{t-1} + \log|\mathbf{V}_t|$
(5) Updating equations	$\mathbf{s}_{(t|t)} = \mathbf{s}_{(t|t-1)} + \mathbf{P}_{(t|t-1)}\mathbf{H}_t^T\mathbf{V}_t^{-1}\mathbf{e}_t$ $\mathbf{P}_{(t|t)} = \mathbf{P}_{(t|t-1)} - \mathbf{P}_{(t|t-1)}\mathbf{H}_t^T\mathbf{V}_t^{-1}\mathbf{H}_t\mathbf{P}_{(t|t-1)}$

Step 5 [Updating equations] Update the estimate of the state vector and its variance covariance matrix,

$$\mathbf{s}_{(t|t)} = \mathbf{s}_{(t|t-1)} + \mathbf{P}_{(t|t-1)}\mathbf{H}_t^T\mathbf{V}_t^{-1}\mathbf{e}_t, \tag{6.1.11}$$

$$\mathbf{P}_{(t|t)} = \mathbf{P}_{(t|t-1)} - \mathbf{P}_{(t|t-1)}\mathbf{H}_t^T\mathbf{V}_t^{-1}\mathbf{H}_t\mathbf{P}_{(t|t-1)}, \tag{6.1.12}$$

where $\mathbf{P}_{(t|t-1)}\mathbf{H}_t^T\mathbf{V}_t^{-1} \equiv \mathbf{K}_t$ is the Kalman gain. When $\mathbf{f}_t^{(o)}$ is omitted from the observation equation, the estimate $\mathbf{s}_{(t|t)}$ is the weighted sum of the observed $\mathbf{Y}_t$ and predicted $\mathbf{s}_{(t|t-1)}$,

$$\mathbf{s}_{(t|t)} = \mathbf{K}_t\mathbf{Y}_t + (\mathbf{I} - \mathbf{K}_t\mathbf{H}_t)\mathbf{s}_{(t|t-1)}, \tag{6.1.13}$$

where $\mathbf{I}$ is the identity matrix. This is the end of the steps.

6.2 Longitudinal Data

6.2.1 State Space Representations of Longitudinal Data

Longitudinal data are composed of values for multiple subjects $i = 1, \cdots, N$ over multiple time points $t = 0, 1, 2, \cdots, T_i$ or $t = 1, 2, \cdots, T_i$. The state space representation of longitudinal data is mostly the same as the state space representation of time series data in Sect. 6.1.1 except for the subscript i. Longitudinal data are represented by a state space defined by two equations: a state equation,

$$\mathbf{s}_{i(t)} = \mathbf{\Phi}_{i(t:t-1)}\mathbf{s}_{i(t-1)} + \mathbf{f}_{i,t}^{(s)} + \mathbf{v}_{i,t}, \tag{6.2.1}$$

and an observation equation,

$$\mathbf{Y}_{i,t} = \mathbf{H}_{i,t}\mathbf{s}_{i(t)} + \mathbf{f}_{i,t}^{(o)} + \mathbf{\xi}_{i,t}. \tag{6.2.2}$$

In the state equation, $\mathbf{s}_{i(t)}$ is a state vector of the ith subject at time t, $\mathbf{\Phi}_{i(t:t-1)}$ is a state transition matrix from $t - 1$ to t, $\mathbf{f}_{i,t}^{(s)}$ is a non-random input vector, and $\mathbf{v}_{i,t}$ is a random input vector with a variance covariance matrix $\mathbf{Q}_{i,t} \equiv \mathrm{Var}(\mathbf{v}_{i,t})$. In the observation equation, $\mathbf{Y}_{i,t}$ is the response vector, $\mathbf{H}_{i,t}$ indicates which elements of the state vector are observed, $\mathbf{f}_{i,t}^{(o)}$ is a non-random input vector, and $\mathbf{\xi}_{i,t}$ is a random input vector with a variance covariance matrix $\mathbf{r}_{i,t} \equiv \mathrm{Var}(\mathbf{\xi}_{i,t})$. The Kalman filter defined in the following section is a recursive algorithm that produces linear estimators of the state vector. The notations $\mathbf{s}_{i(t|t-1)}$ and $\mathbf{s}_{i(t|t)}$ denote the estimate of the state at time t given the observations up to times $t - 1$ and t, respectively. The variance covariance matrices of $\mathbf{s}_{i(t|t-1)}$ and $\mathbf{s}_{i(t|t)}$ are $\mathbf{P}_{i(t|t-1)} \equiv \mathrm{Var}(\mathbf{s}_{i(t|t-1)})$ and $\mathbf{P}_{i(t|t)} \equiv \mathrm{Var}(\mathbf{s}_{i(t|t)})$, respectively.

Table 6.3 State space representations of longitudinal data

	Equations and initial state	Variance
State equation	$\mathbf{s}_{i(t)} = \boldsymbol{\Phi}_{i(t:t-1)}\mathbf{s}_{i(t-1)} + \mathbf{f}_{i,t}^{(s)} + \boldsymbol{\upsilon}_{i,t}$	$\mathbf{Q}_{i,t} \equiv \mathrm{Var}(\boldsymbol{\upsilon}_{i,t})$
Observation equation	$\mathbf{Y}_{i,t} = \mathbf{H}_{i,t}\mathbf{s}_{i(t)} + \mathbf{f}_{i,t}^{(o)} + \boldsymbol{\xi}_{i,t}$	$\mathbf{r}_{i,t} \equiv \mathrm{Var}(\boldsymbol{\xi}_{i,t})$
Initial state	$\mathbf{s}_{i(0\vert 0)}$ or $\mathbf{s}_{i(-1\vert-1)}$	$\mathbf{P}_{i(0\vert0)} \equiv \mathrm{Var}(\mathbf{s}_{i(0\vert0)})$ or $\mathbf{P}_{i(-1\vert-1)} \equiv \mathrm{Var}(\mathbf{s}_{i(-1\vert-1)})$

The state space representation of a model is defined by the state equation with the variance covariance matrix $\mathbf{Q}_{i,t}$, the observation equation with the variance covariance matrix $\mathbf{r}_{i,t}$, and the initial state with the variance covariance matrix, which are specified as shown in Table 6.3. When observations start from $Y_{i,1}$, the initial state is $\mathbf{s}_{i(0\vert0)}$. The observation in autoregressive linear mixed effects models start from $Y_{i,0}$. In this case, the initial state is $\mathbf{s}_{i(-1\vert-1)}$.

6.2.2   Calculations of Likelihoods

For analysis of longitudinal data using linear mixed effects models or autoregressive linear mixed effects models, the maximum likelihood (ML) estimates are obtained by applying an optimization method to minimize $-2ll$. We can obtain closed forms for the ML estimators (MLEs) of the fixed effects parameters if the variance covariance parameters and the autoregressive parameter are given. The fixed effects are then concentrated out of $-2ll$. The marginal and autoregressive forms of $-2ll$ for autoregressive linear mixed effects models are defined in Sects. 2.5.1 and 2.5.2, respectively, where the marginal form is unconditional on the previous response. The marginal form of the model addresses the problem of intermittent missing values because it can be used even if there are missing values in the previous response as a covariate. $-2ll$ for linear mixed effects models is defined in Sect. 1.5.1. It is usually necessary to use matrices whose sizes depend on the number of observations of a subject in likelihood calculations.

The state space representations and the Kalman filter provide us with an alternative method for calculating $-2ll$ of linear mixed effects models (Jones and Ackerson 1990; Jones and Boadi-Boateng 1991; Jones 1993). The Kalman theory assumes that the values of parameters are known (Kalman 1960). The Kalman filter is modified to concentrate the fixed effects out of $-2ll$ (Jones 1986). In this modified method, the filter is applied not only to the observation vector $\mathbf{Y}_i$ but also to each column of the fixed effects design matrix $\mathbf{X}_i$ (Jones 1986, 1993; Jones and Ackerson 1990; Jones and Boadi-Boateng 1991).

The state space representations and the modified Kalman filter are also used (Funatogawa and Funatogawa 2008, 2012) in an autoregressive linear mixed effects model (Funatogawa et al. 2007, 2008a; Funatogawa et al. 2008b). The concentrated $-2ll$ is calculated by applying the filter to $\mathbf{Y}_i$ and $(\mathbf{I}_i - \rho \mathbf{F}_i)^{-1} \mathbf{X}_i$.

The Kalman filter is one way of Cholesky decompositions (Jones 1986, 1993). The inverse of the variance covariance matrix $\mathbf{V}_i^{-1}$ or $\mathbf{\Sigma}_i^{-1}$ has a unique factorization $\mathbf{V}_i^{-1} = \mathbf{L}^T \mathbf{L}$, where $\mathbf{L}$ is a lower triangular matrix. This factorization is called the reverse Cholesky decomposition. Pre-multiplying the model by the matrix $\mathbf{L}$ represents the steps of the Kalman filter (Jones 1986). We do not need a closed-form expression for $\mathbf{L}$ because it will be generated recursively.

This method is suitable for datasets with large numbers of observations of a subject. In the case of large time points or multivariate longitudinal data, these become large. Funatogawa and Funatogawa (2012) used the state space representation and the modified Kalman filter to analyze unequally spaced longitudinal data. The data are treated as being equally spaced with large time points and intermittent missing values by selecting a sufficiently small time unit. In Sect. 3.3, this method is applied.

6.3 Autoregressive Linear Mixed Effects Models

6.3.1 State Space Representations of Autoregressive Linear Mixed Effects Models

We consider a state space representation of the autoregressive linear mixed effects model defined in Sect. 2.3 with an AR(1) error and a measurement error (2.4.1). The model is

$$\begin{cases} Y_{i,0} = \mathbf{X}_{i,0}\boldsymbol{\beta} + \mathbf{Z}_{i,0}\mathbf{b}_i + \varepsilon_{(ME)i,0} \\ Y_{i,t} = \rho Y_{i,t-1} + \mathbf{X}_{i,t}\boldsymbol{\beta} + \mathbf{Z}_{i,t}\mathbf{b}_i + \varepsilon_{(AR)i,t} + \varepsilon_{(ME)i,t} - \rho\varepsilon_{(ME)i,t-1}, (t > 0) \end{cases},$$

$$(6.3.1)$$

with $\mathbf{b}_i \sim \text{MVN}(\mathbf{0}, \mathbf{G})$, $\varepsilon_{(AR)i,t} \sim \text{N}(0, \sigma_{AR}^2)$, and $\varepsilon_{(ME)i,t} \sim \text{N}(0, \sigma_{ME}^2)$, where $t = 0, 1, \cdots, T_i$, ρ is an unknown regression coefficient for the previous response, $\boldsymbol{\beta}$ is a $p \times 1$ vector of unknown fixed effects parameters, $\mathbf{X}_{i,t}$ is a known $1 \times p$ design matrix for fixed effects, $\mathbf{b}_i$ is a $q \times 1$ vector of unknown random effects parameters, $\mathbf{Z}_{i,t}$ is a known $1 \times q$ design matrix for random effects, and $\varepsilon_{(AR)i,t}$ and $\varepsilon_{(ME)i,t}$ are random errors. $\varepsilon_{(AR)i,t}$ is an autoregressive error and $\varepsilon_{(ME)i,t}$ is a measurement error. It is assumed that $\mathbf{b}_i$ and $\boldsymbol{\varepsilon}_i$ are both independent across subjects and independently normally distributed with the mean zero vector and variance covariance matrices $\mathbf{G}$ and $\mathbf{R}_i$, respectively. For this model, the state equation, the observation equation, and the initial state are

$$\left\{ \begin{array}{l} \begin{pmatrix} \mu_{i,t} \\ \mathbf{b}_i \end{pmatrix} = \begin{pmatrix} \rho & \mathbf{Z}_{i,t} \\ \mathbf{0}_{q\times 1} & \mathbf{I}_{q\times q} \end{pmatrix} \begin{pmatrix} \mu_{i,t-1} \\ \mathbf{b}_i \end{pmatrix} + \begin{pmatrix} \mathbf{X}_{i,t}\boldsymbol{\beta} \\ \mathbf{0}_{q\times 1} \end{pmatrix} + \begin{pmatrix} \varepsilon_{(AR)i,t} \\ \mathbf{0}_{q\times 1} \end{pmatrix} \\ \\ Y_{i,t} = \begin{pmatrix} 1 & \mathbf{0}_{1\times q} \end{pmatrix} \begin{pmatrix} \mu_{i,t} \\ \mathbf{b}_i \end{pmatrix} + \varepsilon_{(ME)i,t} \\ \\ \mathbf{s}_{i(-1|-1)} = \mathbf{0}_{(1+q)\times 1} \end{array} \right. , \qquad (6.3.2)$$

with variance covariance matrices,

$$\mathbf{Q}_{i,0} \equiv \mathrm{Var}\begin{pmatrix} \varepsilon_{(AR)i,0} \\ \mathbf{0}_{q\times 1} \end{pmatrix} = \mathbf{0}_{(q+1)\times(q+1)}, \ \mathbf{Q}_{i,t} = \begin{pmatrix} \sigma_{AR}^2 & \mathbf{0}_{1\times q} \\ \mathbf{0}_{q\times 1} & \mathbf{0}_{q\times q} \end{pmatrix} \ \text{for } t > 0,$$

$$\mathbf{r}_{i,t} \equiv \mathrm{Var}\big(\varepsilon_{(ME)i,t}\big) = \sigma_{ME}^2, \ \mathbf{P}_{i(-1|-1)} = \begin{pmatrix} 0_{1\times 1} & \mathbf{0}_{1\times q} \\ \mathbf{0}_{q\times 1} & \mathbf{G} \end{pmatrix}. \qquad (6.3.3)$$

Here, $\mathbf{I}_{a\times a}$ denotes an $a \times a$ identity matrix, and $\mathbf{0}_{b\times c}$ denotes a $b \times c$ matrix whose elements are equal to zero. $\mu_{i,t} = Y_{i,t} - \varepsilon_{(ME)i,t}$ is a latent variable for the true value that we would observe if there were no measurement errors. In contrast, $\varepsilon_{(AR)i,t}$ is a random input included in the true process of $\mu_{i,t}$ and influences the later process. We set an initial state and variance covariance matrix for each subject. The initial estimate of the random effects $\mathbf{b}_i$ is $\mathbf{0}_{q\times 1}$ because $\mathbf{b}_i$ is assumed to be normally distributed with the mean zero.

The correspondences between these and the definitions in Sect. 6.2.1 and Table 6.3 are

$$\mathbf{s}_{i(t)} = \begin{pmatrix} \mu_{i,t} \\ \mathbf{b}_i \end{pmatrix}, \ \boldsymbol{\Phi}_{i(t:t-1)} = \begin{pmatrix} \rho & \mathbf{Z}_{i,t} \\ \mathbf{0}_{q\times 1} & \mathbf{I}_{q\times q} \end{pmatrix}, \mathbf{f}_{i,t}^{(s)} = \begin{pmatrix} \mathbf{X}_{i,t}\boldsymbol{\beta} \\ \mathbf{0}_{q\times 1} \end{pmatrix}, \ \upsilon_{i,t} = \begin{pmatrix} \varepsilon_{(AR)i,t} \\ \mathbf{0}_{q\times 1} \end{pmatrix},$$

$$\mathbf{H}_{i,t} = \begin{pmatrix} 1 & \mathbf{0}_{1\times q} \end{pmatrix}, \mathbf{f}_{i,t}^{(o)} = 0, \boldsymbol{\xi}_{i,t} = \varepsilon_{(ME)i,t},$$

$$\mathbf{s}_{i(-1|-1)} = \mathbf{0}_{(1+q)\times 1}, \mathbf{P}_{i(-1|-1)} = \begin{pmatrix} 0_{1\times 1} & \mathbf{0}_{1\times q} \\ \mathbf{0}_{q\times 1} & \mathbf{G} \end{pmatrix}.$$

Next, we consider a specific example of a state space representation of the autoregressive linear mixed effects model with a time-dependent covariate (2.2.8) with an AR(1) error and a measurement error. The model is

$$\left\{ \begin{array}{l} Y_{i,0} = \beta_{base} + b_{base\,i} + \varepsilon_{(ME)i,0} \\ Y_{i,t} = \rho Y_{i,t-1} + (\beta_{int} + b_{int\,i}) + (\beta_{cov} x_{c\,i,t} + b_{cov\,i} z_{c\,i,t}) + \varepsilon_{i,t}, \quad (t > 0), \quad (6.3.4) \\ \varepsilon_{i,t} = \varepsilon_{(AR)i,t} + \varepsilon_{(ME)i,t} - \rho\varepsilon_{(ME)i,t-1}, \quad (t > 0) \end{array} \right.$$

with $\varepsilon_{(AR)i,t} \sim N\left(0, \sigma_{AR}^2\right)$, $\varepsilon_{(ME)i,t} \sim N\left(0, \sigma_{ME}^2\right)$, and

$$
\begin{pmatrix} b_{\text{base}\,i} \\ b_{\text{int}\,i} \\ b_{\text{cov}\,i} \end{pmatrix} \sim \text{MVN} \left\{ \begin{pmatrix} 0 \\ 0 \\ 0 \end{pmatrix} \begin{pmatrix} \sigma_{\text{base}}^2 & \sigma_{\text{base int}} & \sigma_{\text{base cov}} \\ \sigma_{\text{base int}} & \sigma_{\text{int}}^2 & \sigma_{\text{int cov}} \\ \sigma_{\text{base cov}} & \sigma_{\text{int cov}} & \sigma_{\text{cov}}^2 \end{pmatrix} \right\}.
$$

The equations are rewritten using dummy variables,

$$
\begin{aligned}
Y_{i,t} = {} & \rho Y_{i,t-1} + \beta_{\text{base}} x_{b\,i,t} + \beta_{\text{int}} x_{i\,i,t} + \beta_{\text{cov}} x_{c\,i,t} + b_{\text{base}} z_{b\,i,t} + b_{\text{int}\,i} z_{i\,i,t} \\
& + b_{\text{cov}\,i} z_{c\,i,t} + \varepsilon_{i,t}, \quad (t > 0),
\end{aligned} \tag{6.3.5}
$$

where $\left(x_{b\,i,t}\ x_{i\,i,t}\ x_{c\,i,t} \right)$ are $\left(1\ 0\ 0 \right)$ for $t = 0$ and $\left(0\ 1\ x_{c\,i,t} \right)$ for $t > 0$, and $\left(z_{b\,i,t}\ z_{i\,i,t}\ z_{c\,i,t} \right) = \left(x_{b\,i,t}\ x_{i\,i,t}\ x_{c\,i,t} \right)$. If we have four time points, the response vector is

$$
\begin{aligned}
\begin{pmatrix} Y_{i,0} \\ Y_{i,1} \\ Y_{i,2} \\ Y_{i,3} \end{pmatrix} = {} & \rho \begin{pmatrix} 0 \\ Y_{i,0} \\ Y_{i,1} \\ Y_{i,2} \end{pmatrix} + \begin{pmatrix} x_{b\,i,0}\ x_{i\,i,0}\ x_{c\,i,0} \\ x_{b\,i,1}\ x_{i\,i,1}\ x_{c\,i,1} \\ x_{b\,i,2}\ x_{i\,i,2}\ x_{c\,i,2} \\ x_{b\,i,3}\ x_{i\,i,3}\ x_{c\,i,3} \end{pmatrix} \begin{pmatrix} \beta_{\text{base}} \\ \beta_{\text{int}} \\ \beta_{\text{cov}} \end{pmatrix} \\
& + \begin{pmatrix} z_{b\,i,0}\ z_{i\,i,0}\ z_{c\,i,0} \\ z_{b\,i,1}\ z_{i\,i,1}\ z_{c\,i,1} \\ z_{b\,i,2}\ z_{i\,i,2}\ z_{c\,i,2} \\ z_{b\,i,3}\ z_{i\,i,3}\ z_{c\,i,3} \end{pmatrix} \begin{pmatrix} b_{\text{base}\,i} \\ b_{\text{int}\,i} \\ b_{\text{cov}\,i} \end{pmatrix} + \begin{pmatrix} \varepsilon_{(ME)i,0} \\ \varepsilon_{(AR)i,1} + \varepsilon_{(ME)i,1} - \rho\varepsilon_{(ME)i,0} \\ \varepsilon_{(AR)i,2} + \varepsilon_{(ME)i,2} - \rho\varepsilon_{(ME)i,1} \\ \varepsilon_{(AR)i,3} + \varepsilon_{(ME)i,3} - \rho\varepsilon_{(ME)i,2} \end{pmatrix}.
\end{aligned} \tag{6.3.6}
$$

The state equation, the observation equation, and the initial state of this model are

$$
\left\{ \begin{aligned}
& \begin{pmatrix} \mu_{i,t} \\ b_{\text{base}\,i} \\ b_{\text{int}\,i} \\ b_{\text{cov}\,i} \end{pmatrix} = \begin{pmatrix} \rho\ z_{b\,i,t}\ z_{i\,i,t}\ z_{c\,i,t} \\ 0\ \ 1\ \ 0\ \ 0 \\ 0\ \ 0\ \ 1\ \ 0 \\ 0\ \ 0\ \ 0\ \ 1 \end{pmatrix} \begin{pmatrix} \mu_{i,t-1} \\ b_{\text{base}\,i} \\ b_{\text{int}\,i} \\ b_{\text{cov}\,i} \end{pmatrix} \\
& \qquad + \begin{pmatrix} x_{b\,i,t}\beta_{\text{base}} + x_{i\,i,t}\beta_{\text{int}} + x_{c\,i,t}\beta_{\text{cov}} \\ 0 \\ 0 \\ 0 \end{pmatrix} + \begin{pmatrix} \varepsilon_{(AR)i,t} \\ 0 \\ 0 \\ 0 \end{pmatrix}, \\
& Y_{i,t} = \left(1\ 0\ 0\ 0 \right) \left(\mu_{i,t}\ b_{\text{base}\,i}\ b_{\text{int}\,i}\ b_{\text{cov}\,i} \right)^T + \varepsilon_{(ME)i,t} \\
& \mathbf{s}_{i(-1|-1)} = \mathbf{0}_{4\times 1}
\end{aligned} \right. \tag{6.3.7}
$$

with variance covariance matrices,

$$\mathbf{Q}_{i,0} \equiv \mathrm{Var}\left(\left(\varepsilon_{(AR)i,0}\ 0\ 0\ 0\right)^T\right) = \mathbf{0}_{4\times 4},\ \mathbf{Q}_{i,t} = \begin{pmatrix} \sigma_{AR}^2 & 0 & 0 & 0 \\ 0 & 0 & 0 & 0 \\ 0 & 0 & 0 & 0 \\ 0 & 0 & 0 & 0 \end{pmatrix} \text{for } t > 0,$$

$$\mathbf{r}_{i,t} \equiv \mathrm{Var}\left(\varepsilon_{(ME)i,t}\right) = \sigma_{ME}^2,\ \mathbf{P}_{i(-1|-1)} = \begin{pmatrix} 0 & 0 & 0 & 0 \\ 0 & \sigma_{base}^2 & \sigma_{base\ int} & \sigma_{base\ cov} \\ 0 & \sigma_{base\ int} & \sigma_{int}^2 & \sigma_{int\ cov} \\ 0 & \sigma_{base\ cov} & \sigma_{int\ cov} & \sigma_{cov}^2 \end{pmatrix}. \qquad (6.3.8)$$

6.3.2   Steps for Modified Kalman Filter for Autoregressive Linear Mixed Effects Models

The modified Kalman filter is calculated by the following Steps 1 through 6 for each observation. We begin the calculation by applying the steps to the first observation of the first subject, then to each subsequent observation of the first subject, up to the last observation. The steps are then repeated for each observation of each subject until the last observation of the last subject. The fixed effects are concentrated out of $-2ll$ by applying the filter to $\mathbf{Y}_i$ and $(\mathbf{I}_i - \rho\mathbf{F}_i)^{-1}\mathbf{X}_i$. For this procedure, we use a state matrix $\mathbf{S}_{i(t)}$, with dimensions $(1+q) \times (p+1)$ instead of the state vector $\mathbf{s}_{i(t)}$ with dimensions $(1+q) \times 1$. The values of $(\mathbf{I}_i - \rho\mathbf{F}_i)^{-1}\mathbf{X}_i$ at time t are calculated recursively. The initial state matrix is $\mathbf{S}_{i(-1|-1)} = \mathbf{0}_{(1+q)\times(p+1)}$ for each subject, and a variance covariance matrix, $\mathbf{P}_{i(-1|-1)}$, is defined in (6.3.3). The following steps are then applied to every observation.

Step 1 [Prediction Equations] Calculate a one-step prediction of the state matrix,

$$\mathbf{S}_{i(t|t-1)} = \mathbf{\Phi}_{(t;t-1)}\mathbf{S}_{i(t-1|t-1)}. \qquad (6.3.9)$$

Because the fixed effects are concentrated out, we omit the vector $\mathbf{f}_{i,t}^{(s)}$ in the modified method. The variance covariance matrix of this prediction is

$$\mathbf{P}_{i(t|t-1)} = \mathbf{\Phi}_{(t;t-1)}\mathbf{P}_{i(t-1|t-1)}\mathbf{\Phi}_{(t;t-1)}^T + \mathbf{Q}_{i,t}. \qquad (6.3.10)$$

Step 2 The covariate row vector of the fixed effects is

$$\mathbf{X}_{i,t}^* = \rho\mathbf{X}_{i,t-1}^* + \mathbf{X}_{i,t}. \qquad (6.3.11)$$

The initial values for each subject are

$$\mathbf{X}^*_{i,-1} = \mathbf{0}_{1 \times p}.$$ (6.3.12)

This step produces

$$\mathbf{X}^*_{i,t} = \sum_{j=0}^{t} (\rho^{t-j}) \mathbf{X}_{i,j}.$$ (6.3.13)

This step is particular to the modified version of the Kalman filter with autoregressive linear mixed effects models.

Step 3 Predict the next observation,

$$\left[\mathbf{X}^*_{i,(t|t-1)} \ \mathbf{Y}_{i,(t|t-1)} \right] = \mathbf{H}_{i,t} \mathbf{S}_{i(t|t-1)},$$ (6.3.14)

where the notation $\left[\mathbf{A} \ \ \mathbf{B} \right]$ denotes the matrix $\mathbf{A}$ augmented by matrix $\mathbf{B}$. $\mathbf{Y}_{i,(t|t-1)}$ is the predicted value of $\mathbf{Y}_{i,t}$ given the observations up to time $t-1$, and $\mathbf{X}^*_{i,(t|t-1)}$ is used to calculate $-2ll$.

Step 4 Calculate the innovation row vector $\mathbf{e}_{i,t}$, which is the difference between the row vector of $\mathbf{X}^*_{i,t}$ augmented by $\mathbf{Y}_{i,t}$ and the row vector of $\mathbf{X}^*_{i,(t|t-1)}$ augmented by $\mathbf{Y}_{i,(t|t-1)}$,

$$\mathbf{e}_{i,t} = \left[\mathbf{X}^*_{i,t} \ \mathbf{Y}_{i,t} \right] - \left[\mathbf{X}^*_{i,(t|t-1)} \ \mathbf{Y}_{i,(t|t-1)} \right].$$ (6.3.15)

The variance of this innovation is

$$\mathbf{V}_{i,t} = \mathbf{H}_{i,t} \mathbf{P}_{i(t|t-1)} \mathbf{H}^T_{i,t} + \mathbf{r}_{i,t}.$$ (6.3.16)

where $\mathbf{r}_{i,t} = \sigma^2_{\mathrm{ME}}$ is a scalar.

Step 5 Accumulate the following quantities:

$$\mathbf{M}_{i,t} = \mathbf{M}_{i,t-1} + \mathbf{e}^T_{i,t} \mathbf{V}^{-1}_{i,t} \mathbf{e}_{i,t},$$ (6.3.17)

$$\mathrm{DET}_{i,t} = \mathrm{DET}_{i,t-1} + \log|\mathbf{V}_{i,t}|.$$ (6.3.18)

The initial values of $\mathbf{M}_{i,-1}$ and $\mathrm{DET}_{i,-1}$ are $\mathbf{0}_{(p+1) \times (p+1)}$ and 0 for $i = 1$ and $\mathbf{M}_{i-1,T_{i-1}}$ and $\mathrm{DET}_{i-1,T_{i-1}}$ for $i > 1$. The quantities are accumulated over every observation of every subject. The final values are required to calculate $-2ll$.

Step 6 [Updating Equations] Update the estimate of the state vector,

$$\mathbf{S}_{i(t|t)} = \mathbf{S}_{i(t|t-1)} + \mathbf{P}_{i(t|t-1)} \mathbf{H}^T_{i,t} \mathbf{V}^{-1}_{i,t} \mathbf{e}_{i,t}.$$ (6.3.19)

The updated variance covariance matrix of the state is

$$\mathbf{P}_{i(t|t)} = \mathbf{P}_{i(t|t-1)} - \mathbf{P}_{i(t|t-1)}\mathbf{H}_{i,t}^T \mathbf{V}_{i,t}^{-1} \mathbf{H}_{i,t} \mathbf{P}_{i(t|t-1)}. \qquad (6.3.20)$$

This is the end of the steps.

If $Y_{i,t}$ is a missing observation, we skip Steps 3, 4, and 5 and set $\mathbf{S}_{i(t|t)} = \mathbf{S}_{i(t|t-1)}$ and $\mathbf{P}_{i(t|t)} = \mathbf{P}_{i(t|t-1)}$ in Step 6. Now return to Step 1 and proceed to the next observation, repeating until the final observation.

At the end of the data, where $(i, t) = (N, T_N)$, the matrix $\mathbf{M}_{N,T_N}$ is

$$\begin{bmatrix} \sum_{i=1}^{N}\{(\mathbf{I}_i - \rho\mathbf{F}_i)^{-1}\mathbf{X}_i\}^T \mathbf{\Sigma}_i^{-1}(\mathbf{I}_i - \rho\mathbf{F}_i)^{-1}\mathbf{X}_i & \sum_{i=1}^{N}\{(\mathbf{I}_i - \rho\mathbf{F}_i)^{-1}\mathbf{X}_i\}^T \mathbf{\Sigma}_i^{-1}\mathbf{Y}_i \\ \sum_{i=1}^{N}\mathbf{Y}_i^T \mathbf{\Sigma}_i^{-1}(\mathbf{I}_i - \rho\mathbf{F}_i)^{-1}\mathbf{X}_i & \sum_{i=1}^{N}\mathbf{Y}_i^T \mathbf{\Sigma}_i^{-1}\mathbf{Y}_i \end{bmatrix},$$

$$(6.3.21)$$

and

$$\text{DET}_{N,T_N} = \sum_{i=1}^{N} \log|\mathbf{\Sigma}_i|. \qquad (6.3.22)$$

$\mathbf{M}_{N,T_N}$ and DET_{N,T_N} are used to calculate $-2ll$ with the following equation:

$$-2ll = \sum_{i=1}^{N} n_i \log(2\pi) + \sum_{i=1}^{N} \log|\mathbf{\Sigma}_i| + \sum_{i=1}^{N} \mathbf{Y}_i^T \mathbf{\Sigma}_i^{-1}\mathbf{Y}_i$$

$$- \left\{\sum_{i=1}^{N} \mathbf{Y}_i^T \mathbf{\Sigma}_i^{-1}(\mathbf{I}_i - \rho\mathbf{F}_i)^{-1}\mathbf{X}_i\right\}\hat{\boldsymbol{\beta}}, \qquad (6.3.23)$$

with

$$\hat{\boldsymbol{\beta}} = \left[\sum_{i=1}^{N}\{(\mathbf{I}_i - \rho\mathbf{F}_i)^{-1}\mathbf{X}_i\}^T \mathbf{\Sigma}_i^{-1}(\mathbf{I}_i - \rho\mathbf{F}_i)^{-1}\mathbf{X}_i\right]^{-1} \sum_{i=1}^{N}\{(\mathbf{I}_i - \rho\mathbf{F}_i)^{-1}\mathbf{X}_i\}^T \mathbf{\Sigma}_i^{-1}\mathbf{Y}_i.$$

$$(6.3.24)$$

An optimization method is applied to minimize $-2ll$ and obtain the ML estimates of the variance covariance parameters and ρ. The MLEs of the fixed effects, $\hat{\boldsymbol{\beta}}$, are the above equation where $\mathbf{\Sigma}_i$ and ρ are replaced by their ML estimates.

6.3.3 Steps for Calculating Standard Errors and Predicted Values of Random Effects

The standard errors of the ML estimates are derived from the Hessian of the log-likelihood. The fixed effects parameters are included in the log-likelihood calculation to obtain the standard errors. The Hessian can be obtained numerically. The Kalman filter was used to define the log-likelihood. Here, the state matrix $\mathbf{S}_{i(t)}$ is replaced by an original state vector $\mathbf{s}_{i(t)}$, and the steps from Sect. 6.3.2 are modified slightly as follows.

Step 1 [Prediction Equations] The fixed effects are included in the one-step prediction,

$$\mathbf{s}_{i(t|t-1)} = \mathbf{\Phi}_{(t;t-1)}\mathbf{s}_{i(t-1|t-1)} + \mathbf{f}_{i,t}^{(s)}. \tag{6.3.25}$$

Step 2 Skipped

Step 3 The prediction of the next observation is a scalar,

$$Y_{i,(t|t-1)} = \mathbf{H}_{i,t}\mathbf{s}_{i(t|t-1)}. \tag{6.3.26}$$

Step 4 The innovation is a scalar,

$$e_{i,t} = Y_{i,t} - Y_{i,(t|t-1)}. \tag{6.3.27}$$

Step 5 $\mathbf{M}_{i,t}$ is now a scalar,

$$\mathbf{M}_{i,t} = \mathbf{M}_{i,t-1} + e_{i,t}^2 V_{i,t}^{-1}. \tag{6.3.28}$$

The initial values of $\mathbf{M}_{i,-1}$ are 0 for $i = 1$ and $\mathbf{M}_{i-1,T_{i-1}}$ for $i > 1$.

Step 6 [Updating Equations] Update the estimate of the state vector.

After the final observation, $\mathbf{M}_{N,T_N}$ is

$$\mathbf{M}_{N,T_N} = \sum_{i=1}^{N}\left\{\mathbf{Y}_i - (\mathbf{I}_i - \rho\mathbf{F}_i)^{-1}\mathbf{X}_i\boldsymbol{\beta}\right\}^T \boldsymbol{\Sigma}_i^{-1}\left\{\mathbf{Y}_i - (\mathbf{I}_i - \rho\mathbf{F}_i)^{-1}\mathbf{X}_i\boldsymbol{\beta}\right\}. \tag{6.3.29}$$

$-2ll$ is obtained by substituting $\mathbf{M}_{N,T_N}$ and $\mathrm{DET}_{N,T_N} = \sum_{i=1}^{N}\log|\boldsymbol{\Sigma}_i|$ into

$$-2ll = \sum_{i=1}^{N} n_i \log(2\pi) + \sum_{i=1}^{N}\log|\boldsymbol{\Sigma}_i|$$

$$+ \sum_{i=1}^{N}\left\{\mathbf{Y}_i - (\mathbf{I}_i - \rho\mathbf{F}_i)^{-1}\mathbf{X}_i\boldsymbol{\beta}\right\}^T \boldsymbol{\Sigma}_i^{-1}\left\{\mathbf{Y}_i - (\mathbf{I}_i - \rho\mathbf{F}_i)^{-1}\mathbf{X}_i\boldsymbol{\beta}\right\}. \tag{6.3.30}$$

If the state vector includes random effects, $\mathbf{b}_i$, the updating equation $\mathbf{s}_{i(t|t)}$ in Step 6 of the last observation for each subject is the predicted values of the random effects $\hat{\mathbf{b}}_i$.

6.3.4 Another Representation

The state space representation presented in Sect. 6.3.1 provides the marginal likelihood defined in Sect. 2.5.1 and uses the reverse Cholesky decomposition of $\boldsymbol{\Sigma}_i^{-1}$. The autoregressive linear mixed effects model with a stationary AR(1) error, no measurement error, and given $Y_{i,0}$ is represented as follows:

$$
\begin{cases}
\begin{pmatrix} Y_{i,t} \\ \mathbf{b}_i \end{pmatrix} = \begin{pmatrix} \rho & \mathbf{Z}_{i,t} \\ \mathbf{0}_{q\times1} & \mathbf{I}_{q\times q} \end{pmatrix} \begin{pmatrix} Y_{i,t-1} \\ \mathbf{b}_i \end{pmatrix} + \begin{pmatrix} \mathbf{X}_{i,t}\boldsymbol{\beta} \\ \mathbf{0}_{q\times1} \end{pmatrix} + \begin{pmatrix} \varepsilon_{(\mathrm{AR})i,t} \\ \mathbf{0}_{q\times1} \end{pmatrix} \\[2ex]
Y_{i,t} = \begin{pmatrix} 1 & \mathbf{0}_{1\times q} \end{pmatrix} \begin{pmatrix} Y_{i,t} \\ \mathbf{b}_i \end{pmatrix} \\[2ex]
\mathbf{s}_{i(-1|-1)} = \mathbf{0}_{(1+q)\times1}
\end{cases}
,
$$

with variance covariance matrices,

$$
\mathbf{Q}_{i,t} = \mathrm{Var}\begin{pmatrix} \varepsilon_{(\mathrm{AR})i,t} \\ \mathbf{0}_{q\times1} \end{pmatrix} = \begin{pmatrix} \sigma_{\mathrm{AR}}^2 & \mathbf{0}_{1\times q} \\ \mathbf{0}_{q\times1} & \mathbf{0}_{q\times q} \end{pmatrix}, \ \mathbf{P}_{i(0|0)} = \begin{pmatrix} \sigma_{\mathrm{AR}}^2\left(1-\rho^2\right)^{-1} & \mathbf{0}_{1\times q} \\ \mathbf{0}_{q\times1} & \mathbf{G} \end{pmatrix}.
$$

In this form, $Y_{i,t}$ itself is included in the state vector. This representation provides the conditional likelihood given the previous response defined in Sect. 2.5.2 and uses the reverse Cholesky decomposition of $\mathbf{V}_i^{-1}$.

6.4 Multivariate Autoregressive Linear Mixed Effects Models

This section presents a state space representation of the bivariate autoregressive linear mixed effects model defined in Chap. 4. We consider the following model:

$$
\begin{cases}
\mathbf{Y}_{i,0} = \mathbf{X}_{i,0}\boldsymbol{\beta} + \mathbf{Z}_{i,0}\mathbf{b}_i + \boldsymbol{\varepsilon}_{(\mathrm{ME})i,0} \\
\mathbf{Y}_{i,t} = \rho\mathbf{Y}_{i,t-1} + \mathbf{X}_{i,t}\boldsymbol{\beta} + \mathbf{Z}_{i,t}\mathbf{b}_i + \boldsymbol{\varepsilon}_{(\mathrm{AR})i,t} + \boldsymbol{\varepsilon}_{(\mathrm{ME})i,t} - \rho\boldsymbol{\varepsilon}_{(\mathrm{ME})i,t-1}, \ (t>0)
\end{cases}
.
$$

$$(6.4.1)$$

with $\mathbf{b}_i \sim \text{MVN}(\mathbf{0}, \mathbf{G})$, $\boldsymbol{\varepsilon}_{(\text{AR})i,t} \sim \text{MVN}(\mathbf{0}, \mathbf{r}_{\text{AR}})$, and $\boldsymbol{\varepsilon}_{(\text{ME})i,t} \sim \text{MVN}(\mathbf{0}, \mathbf{r}_{\text{ME}})$. The state equation, the observation equation, and the initial state of this model are

$$
\begin{cases}
\begin{pmatrix} \boldsymbol{\mu}_{i,t} \\ \mathbf{b}_i \end{pmatrix} = \begin{pmatrix} \boldsymbol{\rho} & \mathbf{Z}_{i,t} \\ \mathbf{0}_{q\times2} & \mathbf{I}_{q\times q} \end{pmatrix} \begin{pmatrix} \boldsymbol{\mu}_{i,t-1} \\ \mathbf{b}_i \end{pmatrix} + \begin{pmatrix} \mathbf{X}_{i,t}\boldsymbol{\beta} \\ \mathbf{0}_{q\times1} \end{pmatrix} + \begin{pmatrix} \boldsymbol{\varepsilon}_{(\text{AR})i,t} \\ \mathbf{0}_{q\times1} \end{pmatrix} \\
\mathbf{Y}_{i,t} = \begin{pmatrix} \mathbf{I}_{2\times2} & \mathbf{0}_{2\times q} \end{pmatrix} \begin{pmatrix} \boldsymbol{\mu}_{i,t} \\ \mathbf{b}_i \end{pmatrix} + \boldsymbol{\varepsilon}_{(\text{ME})i,t} \\
\mathbf{s}_{i(-1|-1)} = \mathbf{0}_{(2+q)\times1}
\end{cases}
\tag{6.4.2}
$$

with variance covariance matrices,

$$
\mathbf{Q}_{i,0} \equiv \text{Var}\left(\begin{pmatrix} \boldsymbol{\varepsilon}_{(\text{AR})i,0} \\ \mathbf{0}_{q\times1} \end{pmatrix} \right) = \mathbf{0}_{(q+2)\times(q+2)}, \quad \mathbf{Q}_{i,t} = \begin{pmatrix} \mathbf{r}_{\text{AR}} & \mathbf{0}_{2\times q} \\ \mathbf{0}_{q\times2} & \mathbf{0}_{q\times q} \end{pmatrix} \quad \text{for } t > 0,
$$

$$
\mathbf{r}_{i,t} \equiv \text{Var}\big(\boldsymbol{\varepsilon}_{(\text{ME})i,t}\big) = \mathbf{r}_{\text{ME}}, \quad \mathbf{P}_{i(-1|-1)} = \begin{pmatrix} \mathbf{0}_{2\times2} & \mathbf{0}_{2\times q} \\ \mathbf{0}_{q\times2} & \mathbf{G} \end{pmatrix}.
\tag{6.4.3}
$$

The relationship between $\boldsymbol{\mu}_{i,t}$ in the state equation and the response vector is $\boldsymbol{\mu}_{i,t} = \mathbf{Y}_{i,t} - \boldsymbol{\varepsilon}_{(\text{ME})i,t}$,

$$
\begin{pmatrix} \mu_{1i,t} \\ \mu_{2i,t} \end{pmatrix} = \begin{pmatrix} Y_{1i,t} - \varepsilon_{(\text{ME})1i,t} \\ Y_{2i,t} - \varepsilon_{(\text{ME})2i,t} \end{pmatrix}.
\tag{6.4.4}
$$

$\boldsymbol{\mu}_{i,t}$ is a latent variable for the values that we would observe if there were no measurement errors.

The correspondences between these and the definitions in Sect. 6.2.1 and Table 6.3 are

$$
\mathbf{s}_{i(t)} = \begin{pmatrix} \boldsymbol{\mu}_{i,t} \\ \mathbf{b}_i \end{pmatrix}, \ \boldsymbol{\Phi}_{i(t:t-1)} = \begin{pmatrix} \boldsymbol{\rho} & \mathbf{Z}_{i,t} \\ \mathbf{0}_{q\times2} & \mathbf{I}_{q\times q} \end{pmatrix}, \ \mathbf{f}_{i,t}^{(s)} = \begin{pmatrix} \mathbf{X}_{i,t}\boldsymbol{\beta} \\ \mathbf{0}_{q\times1} \end{pmatrix}, \ \boldsymbol{\upsilon}_{i,t} = \begin{pmatrix} \boldsymbol{\varepsilon}_{(\text{AR})i,t} \\ \mathbf{0}_{q\times1} \end{pmatrix},
$$

$$
\mathbf{H}_{i,t} = \begin{pmatrix} \mathbf{I}_{2\times2} & \mathbf{0}_{2\times q} \end{pmatrix}, \ \mathbf{f}_{i,t}^{(o)} = \mathbf{0}, \ \boldsymbol{\xi}_{i,t} = \boldsymbol{\varepsilon}_{(\text{ME})i,t}.
$$

Random effect $\mathbf{b}_i$ is assumed to follow a q variate normal distribution with the mean vector $\mathbf{0}$. The initial estimate of random effect $\mathbf{b}_i$ is $\mathbf{0}_{q\times1}$.

The following equations are a specific example of the state space representation, corresponding to the model (4.3.1). The state equation and the observation equation are

$$
\begin{pmatrix}
\mu_{1i,t} \\
\mu_{2i,t} \\
b_{1\,\text{base}\,i} \\
b_{1\,\text{int}\,i} \\
b_{1\,\text{cov}\,i} \\
b_{2\,\text{base}\,i} \\
b_{2\,\text{int}\,i} \\
b_{2\,\text{cov}\,i}
\end{pmatrix}
=
\begin{pmatrix}
\rho_{11} & \rho_{12} & z_{b\,i,t} & z_{i\,i,t} & z_{c\,i,t} & 0 & 0 & 0 \\
\rho_{21} & \rho_{22} & 0 & 0 & 0 & z_{b\,i,t} & z_{i\,i,t} & z_{c\,i,t} \\
0 & 0 & 1 & 0 & 0 & 0 & 0 & 0 \\
0 & 0 & 0 & 1 & 0 & 0 & 0 & 0 \\
0 & 0 & 0 & 0 & 1 & 0 & 0 & 0 \\
0 & 0 & 0 & 0 & 0 & 1 & 0 & 0 \\
0 & 0 & 0 & 0 & 0 & 0 & 1 & 0 \\
0 & 0 & 0 & 0 & 0 & 0 & 0 & 1
\end{pmatrix}
\begin{pmatrix}
\mu_{1i,t-1} \\
\mu_{2i,t-1} \\
b_{1\,\text{base}\,i} \\
b_{1\,\text{int}\,i} \\
b_{1\,\text{cov}\,i} \\
b_{2\,\text{base}\,i} \\
b_{2\,\text{int}\,i} \\
b_{2\,\text{cov}\,i}
\end{pmatrix}
$$

$$
+
\begin{pmatrix}
x_{b\,i,t}\beta_{1\,\text{base}} + x_{i\,i,t}\beta_{1\,\text{int}} + x_{c\,i,t}\beta_{1\,\text{cov}} \\
x_{b\,i,t}\beta_{2\,\text{base}} + x_{i\,i,t}\beta_{2\,\text{int}} + x_{c\,i,t}\beta_{2\,\text{cov}} \\
0 \\
0 \\
0 \\
0 \\
0 \\
0
\end{pmatrix}
+
\begin{pmatrix}
\varepsilon_{(\text{AR})1i,t} \\
\varepsilon_{(\text{AR})2i,t} \\
0 \\
0 \\
0 \\
0 \\
0 \\
0
\end{pmatrix},
\qquad (6.4.5)
$$

$$
\begin{pmatrix} Y_{1i,t} \\ Y_{2i,t} \end{pmatrix}
=
\begin{pmatrix}
1 & 0 & 0 & 0 & 0 & 0 & 0 & 0 \\
0 & 1 & 0 & 0 & 0 & 0 & 0 & 0
\end{pmatrix}
\begin{pmatrix}
\mu_{1i,t} \\
\mu_{2i,t} \\
b_{1\,\text{base}\,i} \\
b_{1\,\text{int}\,i} \\
b_{1\,\text{cov}\,i} \\
b_{2\,\text{base}\,i} \\
b_{2\,\text{int}\,i} \\
b_{2\,\text{cov}\,i}
\end{pmatrix}
+
\begin{pmatrix}
\varepsilon_{(\text{ME})1i,t} \\
\varepsilon_{(\text{ME})2i,t}
\end{pmatrix}.
\qquad (6.4.6)
$$

Here, $z_{b\,i,t}$ and $x_{b\,i,t}$ are 1 for $t = 0$ and 0 for $t \neq 0$. $z_{i\,i,t}$ and $x_{i\,i,t}$ are 0 for $t = 0$ and 1 for $t \neq 0$.

6.5 Linear Mixed Effects Models

6.5.1 State Space Representations of Linear Mixed Effects Models

The main theme of this book is autoregressive linear mixed effects models, but we also briefly present the state space representations of linear mixed effects models described in Chap. 1. First, we consider the following linear mixed effects model with a stationary AR(1) error:

$$\begin{cases} Y_{i,t} = \mathbf{X}_{i,t}\boldsymbol{\beta} + \mathbf{Z}_{i,t}\mathbf{b}_i + \varepsilon_{e(AR)i,t} \\ \varepsilon_{e(AR)i,t} = \rho\varepsilon_{e(AR)i,t-1} + \eta_{(AR)i,t} \end{cases}. \tag{6.5.1}$$

with $\mathbf{b}_i \sim \text{MVN}(\mathbf{0}, \mathbf{G})$, $\eta_{(AR)i,t} \sim \text{N}(0, \sigma_{AR}^2)$, and $\varepsilon_{e(AR)i,0} \sim \text{N}\left(0, \sigma_{AR}^2(1-\rho^2)^{-1}\right)$. A stationary AR(1) error is discussed in Sects. 2.4.1 and 2.6. The state equation, the observation equation, and the initial state of this model are

$$\begin{cases} \begin{pmatrix} \varepsilon_{e(AR)i,t} \\ \mathbf{b}_i \end{pmatrix} = \begin{pmatrix} \rho & \mathbf{0}_{1 \times q} \\ \mathbf{0}_{q \times 1} & \mathbf{I}_{q \times q} \end{pmatrix} \begin{pmatrix} \varepsilon_{e(AR)i,t-1} \\ \mathbf{b}_i \end{pmatrix} + \begin{pmatrix} \eta_{(AR)i,t} \\ \mathbf{0}_{q \times 1} \end{pmatrix} \\ Y_{i,t} = \begin{pmatrix} 1 & \mathbf{Z}_{i,t} \end{pmatrix} \begin{pmatrix} \varepsilon_{e(AR)i,t} \\ \mathbf{b}_i \end{pmatrix} + \mathbf{X}_{i,t}\boldsymbol{\beta} \\ \mathbf{s}_{i(0|0)} = \mathbf{0}_{(1+q) \times 1} \end{cases}, \tag{6.5.2}$$

with variance covariance matrices,

$$\mathbf{Q}_{i,t} = \text{Var}\begin{pmatrix} \eta_{(AR)i,t} \\ \mathbf{0}_{q \times 1} \end{pmatrix} = \begin{pmatrix} \sigma_{AR}^2 & \mathbf{0}_{1 \times q} \\ \mathbf{0}_{q \times 1} & \mathbf{0}_{q \times q} \end{pmatrix}, \; \mathbf{P}_{i(0|0)} = \begin{pmatrix} \sigma_{AR}^2(1-\rho^2)^{-1} & \mathbf{0}_{1 \times q} \\ \mathbf{0}_{q \times 1} & \mathbf{G} \end{pmatrix}. \tag{6.5.3}$$

There is no random input to the observation equation.

Next, we consider the following linear mixed effects model with a measurement error:

$$Y_{i,t} = \mathbf{X}_{i,t}\boldsymbol{\beta} + \mathbf{Z}_{i,t}\mathbf{b}_i + \varepsilon_{(ME)i,t}. \tag{6.5.4}$$

with $\mathbf{b}_i \sim \text{MVN}(\mathbf{0}, \mathbf{G})$ and $\varepsilon_{(ME)i,t} \sim \text{N}(0, \sigma_{ME}^2)$. The state equation, the observation equation, and the initial state of this model are

$$\begin{cases} \mathbf{b}_i = \mathbf{b}_i \\ Y_{i,t} = \mathbf{Z}_{i,t}\mathbf{b}_i + \mathbf{X}_{i,t}\boldsymbol{\beta} + \varepsilon_{(ME)i,t} \\ \mathbf{s}_{i(0|0)} = \mathbf{0}_{q \times 1} \end{cases}, \tag{6.5.5}$$

with variance covariance matrices,

$$\mathbf{r}_{i,t} \equiv \text{Var}\left(\varepsilon_{(ME)i,t}\right) = \sigma_{ME}^2, \; \mathbf{P}_{i(0|0)} = \mathbf{G}. \tag{6.5.6}$$

There is no random input to the state equation. Another state space representation of this model can be constructed. The state equation, the observation equation, and the initial state of this representation are

$$
\left\{
\begin{aligned}
\begin{pmatrix} \mathcal{E}_{(ME)i,t} \\ \mathbf{b}_i \end{pmatrix} &= \begin{pmatrix} 0 & \mathbf{0}_{1 \times q} \\ \mathbf{0}_{q \times 1} & \mathbf{I}_{q \times q} \end{pmatrix} \begin{pmatrix} \mathcal{E}_{(ME)i,t-1} \\ \mathbf{b}_i \end{pmatrix} + \begin{pmatrix} \mathcal{E}_{(ME)i,t} \\ \mathbf{0}_{q \times 1} \end{pmatrix} \\
Y_{i,t} &= \begin{pmatrix} 1 & \mathbf{Z}_{i,t} \end{pmatrix} \begin{pmatrix} \mathcal{E}_{(ME)i,t} \\ \mathbf{b}_i \end{pmatrix} + \mathbf{X}_{i,t} \boldsymbol{\beta} \\
\mathbf{s}_{i(0|0)} &= \mathbf{0}_{(1+q) \times 1}
\end{aligned}
\right. \tag{6.5.7}
$$

with variance covariance matrices,

$$
\mathbf{Q}_{i,t} \equiv \mathrm{Var} \begin{pmatrix} \mathcal{E}_{(ME)i,t} \\ \mathbf{0}_{q \times 1} \end{pmatrix} = \begin{pmatrix} \sigma^2_{ME} & \mathbf{0}_{1 \times q} \\ \mathbf{0}_{q \times 1} & \mathbf{0}_{q \times q} \end{pmatrix}, \quad \mathbf{P}_{i(0|0)} = \begin{pmatrix} 0 & \mathbf{0}_{1 \times q} \\ \mathbf{0}_{q \times 1} & \mathbf{G} \end{pmatrix}. \tag{6.5.8}
$$

There is no random input to the observation equation.

6.5.2 Steps for Modified Kalman Filter

The steps for the modified Kalman filter for linear mixed effects models are almost the same as those used for autoregressive linear mixed models, defined in Sect. 6.3.2. The key differences are that we do not need to calculate $\mathbf{X}^*_{i,t}$ (6.3.11) and we can omit Step 2. We replace $\mathbf{X}^*_{i,(t|t-1)}$ and $\mathbf{X}^*_{i,t}$ in Steps 3 and 4 by $\mathbf{X}_{i,(t|t-1)}$ and $\mathbf{X}_{i,t}$. After the steps have been applied to every observation, the matrix $\mathbf{M}_{N,T_N}$ is

$$
\begin{bmatrix} \sum_{i=1}^{N} \mathbf{X}_i^T \mathbf{V}_i^{-1} \mathbf{X}_i & \sum_{i=1}^{N} \mathbf{X}_i^T \mathbf{V}_i^{-1} \mathbf{Y}_i \\ \sum_{i=1}^{N} \mathbf{Y}_i^T \mathbf{V}_i^{-1} \mathbf{X}_i & \sum_{i=1}^{N} \mathbf{Y}_i^T \mathbf{V}_i^{-1} \mathbf{Y}_i \end{bmatrix}, \tag{6.5.9}
$$

and

$$
\mathrm{DET}_{N,T_N} = \sum_{i=1}^{N} \log|\mathbf{V}_i|. \tag{6.5.10}
$$

We use $\mathbf{M}_{N,T_N}$ and DET_{N,T_N} to calculate $-2ll$,

$$-2ll = \sum_{i=1}^{N} n_i \log(2\pi) + \sum_{i=1}^{N} \log|\mathbf{V}_i| + \sum_{i=1}^{N} \mathbf{Y}_i^T \mathbf{V}_i^{-1} \mathbf{Y}_i - \left(\sum_{i=1}^{N} \mathbf{Y}_i^T \mathbf{V}_i^{-1} \mathbf{X}_i \right) \hat{\boldsymbol{\beta}},$$

(6.5.11)

with

$$\hat{\boldsymbol{\beta}} = \left(\sum_{i=1}^{N} \mathbf{X}_i^T \mathbf{V}_i^{-1} \mathbf{X}_i \right)^{-1} \sum_{i=1}^{N} \mathbf{X}_i^T \mathbf{V}_i^{-1} \mathbf{Y}_i.$$

(6.5.12)

When $\mathbf{V}_i$ is written $\sigma^2 \mathbf{V}_{ci}$, σ^2 is estimated as

$$\hat{\sigma}^2 = \frac{1}{\sum_i n_i} \sum_{i=1}^{N} (\mathbf{Y}_i - \mathbf{X}_i \boldsymbol{\beta})^T \mathbf{V}_{ci}^{-1} (\mathbf{Y}_i - \mathbf{X}_i \boldsymbol{\beta}).$$

(6.5.13)

$\hat{\sigma}^2$ is substituted into $-2ll$ in Sect. 1.5.1 to obtain

$$-2ll = \sum_{i=1}^{N} n_i \log(2\pi) + \sum_{i=1}^{N} n_i \log \hat{\sigma}^2 + \sum_{i=1}^{N} \log|\mathbf{V}_{ci}| + \sum_{i=1}^{N} n_i.$$

(6.5.14)

The modified Kalman filter was used to calculate $-2ll$ in Jones (1993). $\mathbf{V}_i$ and $\mathbf{P}_{i(0|0)}$ were replaced by $\mathbf{V}_{ci}$ and $\mathbf{P}_{ci(0|0)}$ where $\mathbf{P}_{i(0|0)} = \sigma^2 \mathbf{P}_{ci(0|0)}$. The Cholesky factorization of the upper part of $\mathbf{M}_{N,T_N}$,

$$\left[\sum_{i=1}^{N} \mathbf{X}_i^T \mathbf{V}_{ci}^{-1} \mathbf{X}_i \quad \sum_{i=1}^{N} \mathbf{X}_i^T \mathbf{V}_{ci}^{-1} \mathbf{Y}_i \right],$$

(6.5.15)

replaces the matrix by $\begin{bmatrix} \mathbf{T} & \mathbf{b} \end{bmatrix}$, where $\mathbf{T}$ is an upper triangular matrix such that

$$\sum_{i=1}^{N} \mathbf{X}_i^T \mathbf{V}_{ci}^{-1} \mathbf{X}_i = \mathbf{T}^T \mathbf{T},$$

(6.5.16)

and $\mathbf{b}$ is

$$\mathbf{b} = \left(\mathbf{T}^T \right)^{-1} \sum_{i=1}^{N} \mathbf{X}_i^T \mathbf{V}_{ci}^{-1} \mathbf{Y}_i.$$

(6.5.17)

Then, $\mathbf{b}^T \mathbf{b}$ and $\hat{\sigma}^2$ are

$$\mathbf{b}^T \mathbf{b} = \left(\sum_{i=1}^{N} \mathbf{X}_i^T \mathbf{V}_{ci}^{-1} \mathbf{Y}_i \right)^T \left(\sum_{i=1}^{N} \mathbf{X}_i^T \mathbf{V}_{ci}^{-1} \mathbf{X}_i \right)^{-1} \sum_{i=1}^{N} \mathbf{X}_i^T \mathbf{V}_{ci}^{-1} \mathbf{Y}_i, \qquad (6.5.18)$$

$$\hat{\sigma}^2 = \frac{1}{\sum_{i=1}^{N} n_i} \left(\sum_{i=1}^{N} \mathbf{Y}_i^T \mathbf{V}_{ci}^{-1} \mathbf{Y}_i - \mathbf{b}^T \mathbf{b} \right). \qquad (6.5.19)$$

Hence, we obtain $-2ll$ (6.5.14) from $\hat{\sigma}^2$ and $\sum_{i=1}^{N} \log|\mathbf{V}_{ci}|$.

References

Anderson TW, Hsiao C (1982) Formulation and estimation of dynamic models using panel data. J Econom 18:47–82

Funatogawa I, Funatogawa T (2008) State space representation of an autoregressive linear mixed effects model for the analysis of longitudinal data. In: JSM Proceedings, Biometrics Section. American Statistical Association, pp 3057–3062

Funatogawa I, Funatogawa T (2012) An autoregressive linear mixed effects model for the analysis of unequally spaced longitudinal data with dose-modification. Stat Med 31:589–599

Funatogawa I, Funatogawa T, Ohashi Y (2007) An autoregressive linear mixed effects model for the analysis of longitudinal data which show profiles approaching asymptotes. Stat Med 26:2113–2130

Funatogawa I, Funatogawa T, Ohashi Y (2008a) A bivariate autoregressive linear mixed effects model for the analysis of longitudinal data. Stat Med 27:6367–6378

Funatogawa T, Funatogawa I, Takeuchi M (2008b) An autoregressive linear mixed effects model for the analysis of longitudinal data which include dropouts and show profiles approaching asymptotes. Stat Med 27:6351–6366

Harvey AC (1993) Time series models, 2nd edn. The MIT Press

Jones RH (1986) Time series regression with unequally spaced data. J Appl Probab 23A:89–98

Jones RH (1993) Longitudinal data with serial correlation: a state-space approach. Chapman & Hall

Jones RH, Ackerson LM (1990) Serial correlation in unequally spaced longitudinal data. Biometrika 77:721–731

Jones RH, Boadi-Boateng F (1991) Unequally spaced longitudinal data with AR(1) serial correlation. Biometrics 47:161–175

Kalman RE (1960) A new approach to linear filtering and prediction problems. J Basic Eng 82D:35–45

Index

© The Author(s), under exclusive licence to Springer Nature Singapore Pte Ltd. 2018
I. Funatogawa and T. Funatogawa, *Longitudinal Data Analysis*, JSS Research Series
in Statistics, https://doi.org/10.1007/978-981-10-0077-5

Printed in the United States
By Bookmasters